TOUGH PLANTS

FOR

TOUGH PLACES

Sharon Amos

TOUGH PLANTS
FOR
TOUGH PLACES

INVINCIBLE PLANTS FOR EVERY SITUATION

FIREFLY BOOKS

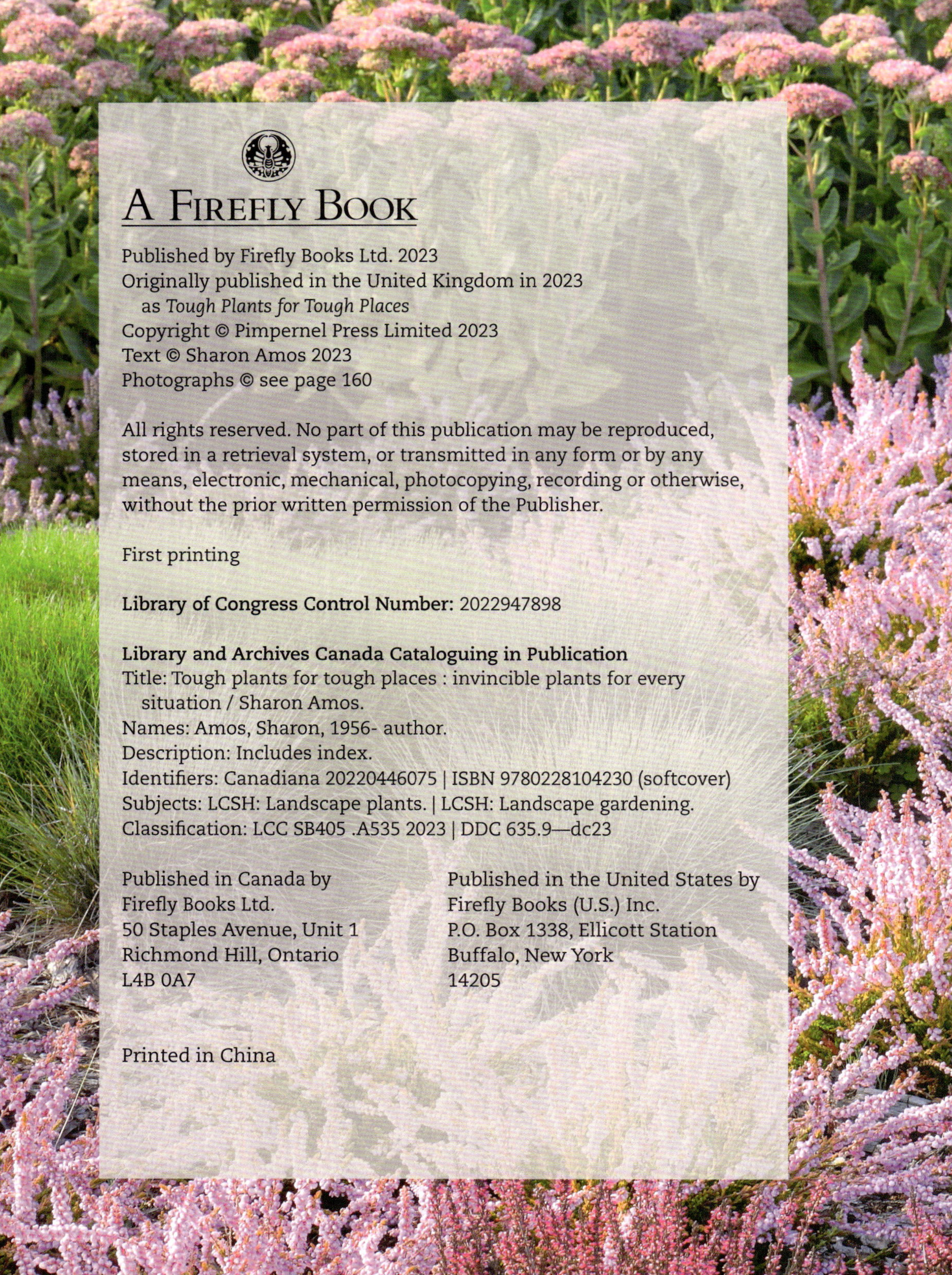

A Firefly Book

Published by Firefly Books Ltd. 2023
Originally published in the United Kingdom in 2023
   as *Tough Plants for Tough Places*
Copyright © Pimpernel Press Limited 2023
Text © Sharon Amos 2023
Photographs © see page 160

First printing

Library of Congress Control Number: 2022947898

Library and Archives Canada Cataloguing in Publication
Title: Tough plants for tough places : invincible plants for every
   situation / Sharon Amos.
Names: Amos, Sharon, 1956- author.
Description: Includes index.
Identifiers: Canadiana 20220446075 | ISBN 9780228104230 (softcover)
Subjects: LCSH: Landscape plants. | LCSH: Landscape gardening.
Classification: LCC SB405 .A535 2023 | DDC 635.9—dc23

Published in Canada by
Firefly Books Ltd.
50 Staples Avenue, Unit 1
Richmond Hill, Ontario
L4B 0A7

Published in the United States by
Firefly Books (U.S.) Inc.
P.O. Box 1338, Ellicott Station
Buffalo, New York
14205

Printed in China

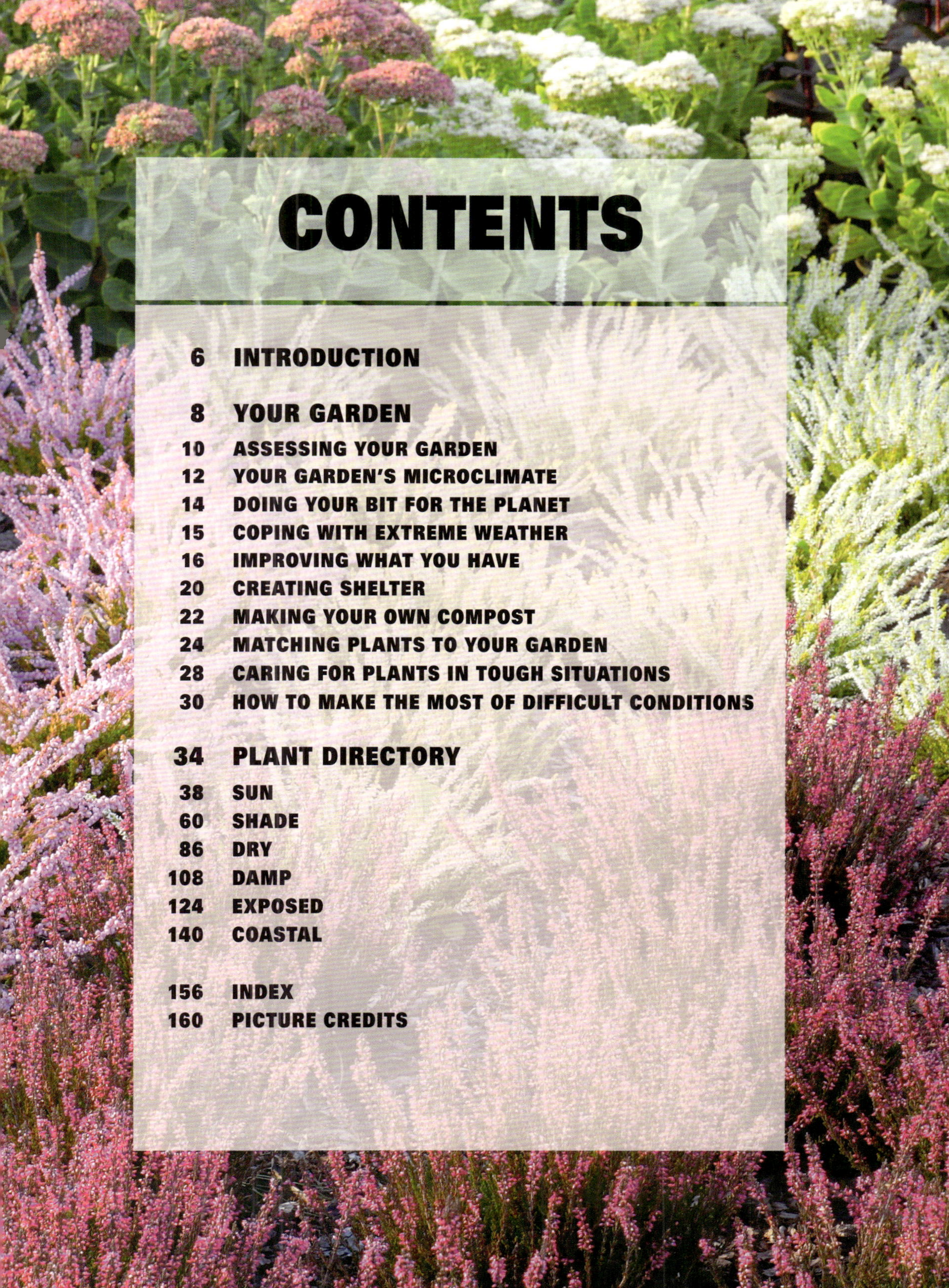

CONTENTS

INTRODUCTION

Most gardens do not have smooth, flat lawns and borders of rich, easily dug soil. We have to put up with damp, sunless corridors between houses, awkward slopes or plots shaded by trees or neighboring buildings. Equally difficult to plant are coastal gardens exposed to gale-force winds and salt spray; waterlogged plots, where the drainage is poor; and dry ground exposed to the glare of the sun day after day, without the slightest shade. In short, few gardens benefit from perfect conditions.

If you try to plant any of these difficult areas with a standard selection of plants from a garden center, the chances are that you will end up with an unsightly display of brown leaves and dried stems, as the plants are defeated by the unsuitable habitat.

What you need for these sites are tough plants that will not only shrug off all the worst conditions in your garden but actually thrive in them and grow well. Of course, these plants are not unkillable but they are practically invincible in the specific hostile conditions they have evolved to cope with. *Tough Plants for Tough Places* is by no means a definitive guide to every plant: rather, use it as a blueprint to get you started on the way to a great garden.

LEFT The tough leathery leaves of shrubby hare's ear, *Bupleurum fruticosum,* can withstand the drying effects of strong winds and harsh sun. Plant it in a coastal garden or on a city balcony.

YOUR
GARDEN

ASSESSING YOUR GARDEN

Before you start to plant, you need to build an accurate picture of your garden's terrain. It's rare that a garden is uniform throughout. Most have one or two problem areas, such as the dense, dry shade beneath an evergreen tree or a patch of poor, stony soil where construction rubble has been dumped in the past.

GETTING TO KNOW YOUR SOIL

The soil is as good a place as any to begin your assessment. Scooping up a handful of soil can tell you a lot about it. Sandy soil will run through your fingers and will have a gritty feel. In prolonged wet weather, clay soil is heavy and sticky, like a handful of modelling clay, and when all the moisture has gone it becomes rock hard. Chalky soil is pale and stony, while peaty soil – the least likely of all garden soils – is brown and wet.

Loam is best of all, a mixture of clay, silt and sand in roughly equal quantities. It is brown and crumbly, without being as sticky as clay or as free-draining as sand. If you squeeze it and it holds its shape for a moment and then crumbles, you have loam. Getting to know your soil will help you to choose plants that will thrive in it.

SOIL PH

The fertility of soil is also directly related to its pH value. This determines whether the soil is acidic, neutral or alkaline (limy). Ideally, soil should be neutral or slightly acidic with a pH value of between 6.5 and 7. (Values below 7 are acidic; those above indicate alkaline soil.) Soils that are too acidic or too alkaline prevent plants from taking up vital nutrients, resulting in poor growth.

PAGE 8–9 Sedums and succulents have thick fleshy leaves that resist drying out and store water.

LEFT Azaleas are small, shrubby rhododendrons that flourish on acidic soil with a burst of spring color. Rhododendrons come in all shapes and sizes, from trees to dwarf alpine species.

SOIL CHECKLIST

SOIL TYPE	CHARACTERISTICS
Sandy	• Contains little organic materials, which means that it tends to be poor in nutrients • Dries out quickly • The gritty, sandy structure makes it free-draining
Clay	• Is heavy and sticky • Will bake solid and crack in summer • May get waterlogged in wet spells • Slow to drain and badly aerated
Peaty	• Acidic • Contains plenty of organic material in a partly rotted state • Wet and may even be waterlogged
Loam	• The easiest soil to work • Rich in organic material and therefore in nutrients • Stays moist without becoming waterlogged
Silt	• Finely grained and fertile • Easily compacted, and once the soil becomes packed down, the air pockets get flattened and the soil drains badly
Chalky	• Alkaline • Tends to be shallow and studded with stones • Dries out quickly • Generally poor in nutrients

Garden centers sell kits and slightly more expensive electronic meters so that you can measure your soil's pH. If you use either of these, take a range of samples from around the garden, because pH can vary considerably within a small area. You can also send soil samples to labs for professional testing.

An easier – and instant – way to assess soil pH is to look at the plants already growing in your own garden and in your neighbors'. Vigorous, healthy rhododendrons and camellias and beds of heather indicate an acidic soil, as these species dislike alkaline or limy conditions. Lilac, hawthorn, fuchsias and achillea (yarrow), on the other hand, may point to an alkaline soil.

USING A SIMPLE SOIL TESTING KIT

Mix a sample of soil with the chemical solution supplied with the test kit, then assess the resulting color once it has had time to develop – usually after a few minutes.

YOUR GARDEN'S MICROCLIMATE

SUN AND SHADE

Sunlight plays an essential role in photosynthesis, the process by which leaves turn carbon dioxide and water into sugars to make the plant grow. But the amount of light that different plants need varies enormously, as can the range of light levels that a particular species will tolerate. Some sun-loving plants will still do well in light shade; and it is possible to grow some shade-loving plants in sunnier positions than normal, usually with the proviso that the soil doesn't dry out.

Trees, bushes, buildings, hedges and fences all cast shade, but rarely do they plunge an area into permanent gloom. You need to walk around the garden at different times of day to see where sunlight and shadows fall. Remember, too, that in winter the sun is much lower in the sky and can reach into more tucked-away corners.

Dappled summer shade from deciduous trees and shrubs can be used to mimic woodland conditions. Evergreen bushes and trees cast a denser shadow, but there are still plants that can cope with that degree of limited light. The darkest areas tend to be found in city gardens, where older houses are close together, creating narrow, shaded corridors. But even these sunless corridors can be brought to life with the right plants.

At the other extreme are open sites, away from trees and buildings, which are exposed to full sun throughout the day.

Early morning sun can be tricky. On frosty winter and spring mornings a blast of sunshine can thaw frozen buds and leaves too quickly, reducing parts of the plants to a pulp.

WIND

Circulating air in a garden is vital for healthy plants. It prevents the build-up of fungal spores and subsequent plant diseases. But you can, of course, have too much of a good thing. Strong winds can cause excessive transpiration in plants, making them lose too much water from the leaf surfaces so that the whole plant wilts and keels over. This effect is particularly noticeable on warm, windy days. Plants usually recover in the cool of the evening, but if they have lost too much water the leaves may show signs of scorching, with blackened edges or dried brown areas. Strong winds also cause permanent damage by snapping fragile stems and by weakening the roots' grip on the soil so that the plant is lifted partly out of the ground – the technical term is wind rock.

Some winds are warmer than others, and you will soon work out which they are. If your garden is in an area dominated by strong, cold winds, you may find scorched leaves on only one side of a plant. On very exposed sites, and particularly in coastal gardens, shrubs and trees will be shaped by the wind into typical bent forms, bowed away from the direction of the strongest gales.

CHECKLIST

Before you can start to make decisions about which plants to buy and where to plant them, you need to get an overall picture of your garden by investigating the following:

- Soil type
- Soil pH
- Shady areas
- Sunny areas
- Exposed areas
- Sheltered areas

ABOVE Many grasses are ideally suited to open positions and hot, sunny gardens. In their natural habitat many species of ornamental grasses often grow in poor soil.

LEFT Shady gardens offer protection from the sun for large-leaved plants. Some variegated plants and shrubs can be damaged by overexposure to sunlight, so they too appreciate a shady corner, while at the same time their leaves can help brighten up the area.

DOING YOUR BIT FOR THE PLANET

Gardens can be an overlooked resource for improving the environment and boosting wildlife. Far from being just a place to grow flowers, with a few simple tweaks to your gardening style they can be home to insects, reptiles, small mammals and birds. By leaving grass to grow long, not rushing to cut down dead stems at the end of summer – in short by not being too tidy – your garden can become a link in the chain of a wider urban wildlife habitat.

Making your own compost from prunings and vegetable scraps from the kitchen reduces transport costs associated with domestic refuse collection and buying-in ready-made soil-improving products and artificial fertilizers. Saving seed from plants that you've grown might seem a small thing but behind every packet of seeds you buy lie the hidden costs of growing, picking, packing and transport. These changes may seem tiny but they make a difference. Remember how back in the 1980s the thought that stopping using hairspray, at least until the use of CFCs in aerosols was banned, seemed such a tiny step to take when set against the ever-increasing hole in the ozone layer? Well, the ozone layer is continuing to recover . . .

Professor Alistair Griffiths, director of science at the Royal Horticultural Society (RHS), is positive that we as gardeners can make a difference to climate resilience overall: 'Gardening is a way the general public can take action to help to mitigate climate change.'

LEFT Common broom, *Cytisus scoparius*, produces masses of small pea-like flowers from April to June. It is a valuable nectar source for bees and is one of the RHS's Plants for Pollinators.

COPING WITH EXTREME WEATHER

No garden stays the same from year to year. Every year gardeners are faced with changes in growth and competition between plants, adopting strategies to make the most of one plant's success and disguise one that's struggling, and this resourcefulness and resilience make gardeners ideally placed to cope with extreme weather events.

The focus of this book is on choosing the right plant for the right place in your garden. But what if nature intervenes? Gardeners have plenty of 'tools' at their fingertips to use in an emergency or to plan longer term.

Creating raised beds to grow plants in gives you much more control. For example, a raised bed of drought-resistant plants in a hot, sunny garden allows water to drain away easily if there is a spell of torrential rain.

Well-mulched soil encourages worms, which in turn aerate the soil and improve its water-holding capacity, reducing run-off and the risk of flooding.

Most gardeners already have a stock of bubble wrap or horticultural fleece to protect borderline hardy plants when a cold snap is forecast; we may have to get used to using them more frequently and for longer.

LEFT Be prepared before cold weather hits and wrap borderline hardy plants in horticultural fleece or bubble wrap. Wrappings need to be transparent to allow light in. Protect the roots of potted plants from freezing by wrapping the pot too.

IMPROVING WHAT YOU HAVE

By choosing tough plants for tough situations, you are working with nature, not against it, but it's always a good idea to give your plants the best possible start. There are plenty of ways to do this, from enhancing the structure of the soil to providing extra protection, to help plants get established.

IMPROVING YOUR SOIL

Soils can be too wet, too dry, too acidic, too alkaline or lacking in nutrients. There are strategies for dealing with all these problems, and throughout this section one 'miracle' solution keeps recurring – well-rotted manure. It can improve drainage in one soil, yet increase moisture-retention in another, lower pH values towards neutral and, of course, add plant nutrients.

Never add fresh manure to the soil. When it rots down it uses up valuable nitrogen, starving the plants it was intended to help.

MAKING CLAY SOIL EASIER TO WORK WITH

Raw, uncultivated clay soil needs extra drainage. Digging in horticultural grit or gravel will help. Add at least a bucketful per square meter (yard) or so and dig it in to the depth of a garden fork or about 25 cm (10 in). Timing is critical. In winter the ground can be frozen or too wet to dig, but in summer it can be so hard that a garden fork will rebound without making an impression. In spring, as soon as the soil is relatively workable – not too hard and not too sticky – get as much work done as you can. Using a fork rather than a spade makes digging clay easier: the sharp edge of a spade can flatten precious air pockets in the soil and actually make drainage worse.

LEFT A thick layer of chipped bark helps soil to retain water and also prevents weeds from growing.

RIGHT Dig in home-made compost to improve water retention in sandy soils or to open up heavy clay or use it as a mulch in spring or autumn.

Well-rotted manure and home-made garden compost will improve clay soil by aerating it and helping it drain better. Again, dig it in when the soil is workable.

To reduce the cracking and baking that are typical of clay soil in summer, use a mulch (see page 30). Try bark or wood chippings or well-rotted manure. Spread the mulch on the soil surface between autumn and spring, but only if the soil is wet after a good downpour. Mulching dry soil will slow down the time the rain takes to reach it and make things worse. Similarly, do not mulch frozen soil, or it will take much longer to thaw.

IMPROVING SANDY AND CHALKY SOILS

Rainwater washes through both these soils quickly, taking with it vital nutrients before the roots have a chance to absorb them. Adding plenty of well-rotted manure or home-made garden compost acts in two ways to improve the soil: it bulks up the soil structure and it provides extra nutrients.

Chalky soil tends to be alkaline. Adding organic materials such as manure and compost, which are naturally acidic, can lower the soil's pH towards neutral, allowing you to grow a greater range of plants. Every time you dig chalky soil, pick out as many stones as you can to give roots room to grow without obstruction.

Both soil types benefit from mulching to conserve moisture. Well-rotted garden compost, bark chippings or even a thick layer of gravel will all help to prevent precious water evaporating from the surface of the soil.

WORKING WITH ACIDIC SOIL

A wide range of plants – rhododendrons, camellias and heathers, for example – will not grow in any soil other than acidic. Even so, soil can be too acidic at the lower end of the scale, and the pH may need to be increased by adding garden lime to bring it up to a more suitable level. The recommended time to do this is in winter: scatter lime on the surface and leave it to be washed into the soil with the rain. The amount you need to add will vary according to soil type and existing pH – somewhere between 100–200 g (3½–7 oz) per square meter (yard). The effect is not instant and is fairly temporary, too, and will have to be repeated every few years. Do not be tempted to sprinkle on a little more 'for luck' – too much lime will affect the nutrients available in the soil just as adversely as over-acidity.

DEALING WITH WATERLOGGED SOIL

If you have soil that stays permanently wet all year-round there is a whole category of exotic species that you can grow. But even water-loving plants need a degree of drainage: digging in well-rotted manure or garden compost opens up air spaces that allow water to pass through.

If water seems to lie temporarily on the surface after rain, the soil could simply have become compacted. This is a typical problem with gardens of newbuild houses where heavy machinery has been at work. All you may need to do is break up the soil with a garden fork to help the water drain away.

ABOVE, TOP Creating a bog garden in a water-logged area means you can grow exotic moisture-loving primulas.

ABOVE Mat-forming sedums can help stabilize soil and are ideal for dry sunny areas.

LEFT Provided soil is acidic, heathers are otherwise undemanding plants.

STABILIZING A BARE SLOPE

The topsoil of a newly created bank or slope is vulnerable. On dry days the wind can simply blow off the surface soil in a cloud of dust, while heavy rain will wash it away. What it needs is a network of specialized plants to hold the surface together and to create a stable environment for other plants to grow in.

Plants that form a flat rosette of leaves are ideal. The leaves radiate from the center and protect the surrounding earth from wind and rain. Verbascums, globe thistles and sea lavender all grow from a basal rosette. Once they have done their job and stabilized the soil, some of them can be divided and moved to create space for other plants.

Mat-forming plants protect the soil, too, but are more difficult to plant into later on, as you need to cut away part of the plant to form a planting space. Nevertheless, they are still worth using and you may find you like the look of a bank and decide to keep it that way.

CREATING SHELTER

MAKING A WINDBREAK

Creating some shelter in an exposed garden will make an enormous difference to the range of plants you can grow, and they will be healthier and more vigorous.

Although it may seem like the ideal solution, surrounding a garden with brick walls or solid fences isn't always the answer. In fact, it can make things worse. When a strong wind hits a solid barrier it actually picks up speed. On the supposedly sheltered side of the wall or fence, it whirls and eddies and can do just as much damage to plants as the full, unfettered force of the wind.

A permeable windbreak that filters the wind and slows it down without any unwanted side effects is best. This can be anything from a length of garden netting temporarily stretched between two canes to a permanently planted hedge. Open-slatted fencing, a chainlink fence or a system of trellis and posts are other options.

Research has shown that a windbreak can shelter land to a distance many times greater than its height. It creates a microclimate – a small area of protected environment within the harsher general climate – and plants growing in this area will grow more vigorously and flower for longer, too.

LEFT A permeable windbreak such as trellis is better than a solid fence at reducing wind speed and protecting plants.

RIGHT An informal hedge will shelter smaller shrubs and plants in front.

PLANTING A HEDGE AS A WINDBREAK

On exposed sites choose the toughest species – hawthorn, blackthorn and sea buckthorn – that can survive gale-force winds. If you're starting from scratch, the small whips or bushes will need some temporary protection until their roots are firmly established. Trellising, chainlink or temporary fencing in place for a couple of years will give them a good start.

BEWARE OF FROST POCKETS

When making a windbreak on a sloping site take care that you do not create a frost pocket. A windbreak can trap cold air and stop it from rolling down a slope or hill. To maintain free passage of air, clear out the bottom of a hedge regularly and make sure that fencing has space at the base for air to circulate.

HEDGES

Hedges make good windbreaks if you have the space: an informal hedge of native bushes can be at least 2 m (6½ ft) thick and a clipped hedge is only slightly less. However, if you are restricted in what you can grow, the sacrifice of space can be well worth it. On exposed sites choose the toughest species – hawthorn, blackthorn, sea buckthorn – that can survive gales of 96 km (60 miles) an hour or searing, salt-laden sea winds. As your hedge will be doing such an important job, treat it well. Prune it at least twice a year: exactly when will depend on the species you have planted. Evergreens, like laurel and some conifers, can be cut back in spring and autumn; box and privet can be trimmed even more regularly throughout the summer for a really neat appearance. Even informal natural hedges benefit from a light trim from time to time to keep them growing strongly.

MAKING YOUR OWN COMPOST

You can never have too much well-rotted garden compost. Dig it into the ground as a soil improver, add a handful to planting holes to help new plants settle in and spread it thickly as a mulch.

Every gardener has favorite tips for making compost, some secret and some not so secret. There is so much written about it and endless information from experts available in magazines, books and on the Internet. You can even take compost-making courses. But there are a few basic principles to follow that should ensure some measure of success, whatever you add.

GOOD COMPOSTING

The best compost comes from keeping a balance between wet and dry ingredients. Whether you build a compost pile or use a manufactured composter, the principles are the same. Build up a heap in layers, using absorbent paper, cardboard and even old rags between layers of wet kitchen scraps, grass cuttings and green, sappy plant cuttings.

Build your heap directly on the ground so that worms and soil organisms can get to work on it. A base layer of twigs or a couple of shallow channels dug into the ground will ensure that air can still circulate throughout the heap. Between additions, cover it with a piece of old carpet or black plastic to keep the warmth in. Once the heap is about a meter (3 feet) high or as tall as you can manage comfortably, weight the cover down with bricks and leave it for a year.

For a home-made composter, use an old garbage bin with the bottom knocked out and some holes drilled in the sides for air circulation. This is a neater way of making compost in a small garden. If you have plenty of space, tie together some old pallets or door panels to make two square heaps side by side. That way you can have one heap under cover, working its magic, while you are building up the next one. Manufactured composters, usually of heavy-duty plastic, work on the same principles. Remember to turn the compost frequently and to keep the lid on tight. As well as garden prunings, you can add some kitchen scraps to your compost heap too (see opposite).

If you don't have space for a compost heap, consider using a wormery to turn your kitchen waste into rich nutritious worm compost. There are plenty of kits on the market that comprise a system of trays that will stack neatly in a corner and include a starter population of red wiggler worms. As well as compost, they also produce a liquid plant feed, usually referred to as 'worm tea'.

Autumn leaves rot by a different process. Make a special heap for them alone and leave them for two years to make your own leaf mold. Twigs take ages to rot and make finished compost awkward to spread or dig into soil. They are only useful as a base layer to allow air into the heap. If you have a shredder, chip them to make a mulch. Even the neatest of compost heaps is not that attractive to look at. If you have space, screen off a small area of your garden with hedging or trellis and climbers, and site your compost heaps there.

It's a good idea to pave the area behind the hedge or screen. With heavy regular use, grass soon gets trampled into mud in winter, especially if you are trundling a wheelbarrow back and forth, for example.

MAKING A COMPOST HEAP

1 If you have space, make two heaps. Use a series of boards that slot into the sides, adding them as the heap gets higher. The heap on the right has rotted down and the resulting compost is being used in the garden.

2 Cover the compost heap with plastic sheeting or with a piece of old carpet to keep the warmth in and help the composting process.

3 You will know when your compost is ready, because well-rotted compost is rich, brown and crumbly and bears no resemblance to the mixture of materials that went into the original heap.

WHAT TO ADD TO YOUR COMPOST HEAP

You can add almost any organic material to a compost heap:

- Raw vegetables and fruit scraps
- Eggshells
- Teabags
- Coffee grounds
- Crumpled newspaper
- Strips of thin cardboard
- Paper towels and tissues
- Torn up rags in natural fibers (cotton or wool, for example)
- Some weeds (only sappy young green leaves). Tie up the worst weeds in a black plastic bag and leave them out in the full sun for a some weeks, or soak them in a bucket of water for a similar time before adding to the compost heat.
- Garden prunings
- Grass cuttings

WHAT NOT TO ADD

- Meat or fish (you will attract rats)
- Bones
- Fats
- Oil
- Scraps of cooked food
- Diseased plant material
- Some weeds (avoid adding weeds that have gone to seed)

MATCHING PLANTS TO YOUR GARDEN

Working with nature is the easiest way to plant a garden. Even a small garden has shady corners and sunny ones. On a larger site there's every likelihood you'll find patches of dry soil and damp areas, slopes, ditches or poor stony soil. There's far less risk of failure if you match plants to these different habitats, because they have evolved over thousands of years to cope with them.

CHOOSING THE RIGHT PLANTS FOR YOUR GARDEN

Once you have become familiar with some of the ways in which plants deal with different conditions, you may even be able to assess an unfamiliar plant and make a reasonable guess where it will be happiest in your garden.

PLANTS THAT THRIVE IN HOT, SUNNY, DRY SITES

One of the key ways that plants deal with drought is by having smaller, narrower leaves. Small leaves, especially thin, needle-like ones, lose far less water than big, flat leaves with a large surface area. The leaves of common broom (*Cytisus scoparius*, page 146) are virtually non-existent, for example, and are no more than tiny scales on its tough, wiry, green stems. Leaves of other drought-tolerant species may have a tough, shiny cuticle to prevent water loss, such as those of shrubby hare's ear (*Bupleurum fruticosum*, page 142).

Many plants of hot dry places have thick, fleshy leaves that can actually store water – the ice plant (*Sedum spectabile*, page 102) is typical.

LEFT Sea holly (*Eryngium bourgatii*) thrives in hot, sunny gardens. The silvery bracts round the flower heads reflect the sun and prevent cell damage.

Plants with scented leaves have developed a novel method of reducing water loss. On hot days the volatile oils that give the plant its characteristic perfume evaporate from the leaves and form a protective 'cloud' around the plant to retain moisture.

A number of factors reduce the likelihood of leaves scorching in the full glare of the sun. Silver leaves reflect light so that it bounces back off the leaf without damaging it. Sometimes it's not the leaves themselves that are silver but the layer of hairs that cover them. In addition to reflecting light, hairs also shade the leaf and reduce water loss and can even combat the effects of salt damage in coastal gardens. The silver-leaved sea holly (*Eryngium bourgatii*, page 147) flourishes in dry conditions – the drier the better – and its leaves turn even more intensely silver as the environment becomes more hostile.

Plants that form a densely packed rosette of leaves – for example, common houseleek (*Sempervivum tectorum*, page 103) – have also adapted to growing in full sun. The overlapping leaves shade each other and reduce water loss.

WINDY, EXPOSED SITES

Plants that have adapted to open, exposed habitats tend to have leaves that are split into small leaflets or that are feathery and finely divided. In a gale these are far less likely to be damaged than a big, flat leaf, which is easily torn.

Habitat also shapes overall appearance. Plants of exposed places tend to form low-growing, rounded shapes that resist the wind, with all their branches and leaves neatly tucked in, so that the wind skims right over. Out of sight, they have deep root systems that anchor them firmly in the earth.

Small species of hebe (*Hebe albicans*, page 133), with their neat, rounded shape, do well on exposed coastal sites; heathers (*Erica cinerea*, pages 128–9,

LEFT Heather species (*Erica* spp.) are low growing and a good choice for exposed windy sites.

and *Calluna vulgaris*) survive by similar means on exposed inland gardens.

Do not assume a plant that grows well in an exposed coastal garden will transplant equally happily to an open, windy site inland. Although coasts may be buffeted by gales, it's always a few degrees warmer in winter because of the warming presence of the sea, which cools down more slowly than the land.

COPING WITH SHADE

All plants need light to photosynthesize, see page 12, but their requirements vary enormously. Even some sun-loving plants will tolerate a surprising degree of shade, although they may flower less prolifically than they would in full sun or produce less upright and more sprawling growth.

BELOW LEFT The spring-flowering wood anemone (*Anemone nemorosa*) is a woodland species that prefers dappled shade.

BELOW RIGHT Variegated shrubs such as elaeagnus brighten up shady corners.

DAMP SHADE

Shade plants that prefer damp soil often have large, thin leaves to catch maximum sunlight and rain. Those that are essentially woodland species grow quickly and flower early in the year before the tree canopy develops. Thereafter, smaller plants take a backseat – indeed, many eventually die back altogether until the following spring, including the wood anemone (*Anemone nemorosa*, page 61) and spring bulbs such as snowdrops (*Galanthus nivalis*) and English bluebells (*Hyacinthoides non-scripta*).

DRY SHADE

The combination of dry soil and shade defeats many plants in the struggle to get enough light while minimizing water loss. But there are some species that can cope. Their leaves often have a shiny surface to reflect light back on to the undersides of neighboring leaves – for example, elaeagnus (*Elaeagnus pungens*, page 128).

GETTING CLUES FROM LATIN NAMES

Latin names for plants can seem off-putting or intimidating. They are often long, complicated, difficult to spell and even more difficult to pronounce. Yet apart from ensuring that you buy exactly the plant you are looking for rather than relying on the common name, which can vary from region to region, Latin names can sometimes help you to find out more about a plant. It's the second word in the two-part name that is the more useful in this respect. The first part of the name denotes the genus to which the plant belongs – a group of fairly similar plants. The second word in the Latin name is the specific name, which can give many clues to a plant's preferred habitat.

As a great part of our own language is derived from Latin, some of these words will seem familiar. You do not need a degree to translate them into some kind of working sense for gardeners.

The words listed to the right are all adjectival, which means that the endings will vary slightly when matched to a genus, to agree with it in gender and in number. Hence sea kale is *Crambe maritima* rather than *maritimus*.

Water-loving
aquaticus
aquatilis
hydrophilus
pluvialis

Dry-loving
aridus
xerophilus

Sun-loving
apricus
solaris

Cold-tolerant
algidus
frigidus

Woodland
sylvicola
sylvatica

Seashore
maritimus
littoralis
marinus

Meadow
arvensis

Marsh
palustris

Sandy soil
arenarius
arenicola

Chalky soil
calcicola

Acid soil
calcifugus

BELOW Forget-me-not (*Myosotis sylvatica*) is a woodland species in the wild.

CARING FOR PLANTS IN TOUGH SITUATIONS

All the plants in the directory (pages 34–155) have been chosen for their ability to grow in tough situations, but when you are introducing small, new plants into the garden, this doesn't mean that you can simply put them in the ground and forget about them.

GIVING PLANTS THE BEST

It's always best to start with relatively small plants that will adapt to the specific conditions in your particular garden as they grow and establish. But small plants are also the most vulnerable and do need a helping hand. If you invest time and effort in giving them a good start when you plant and a bit of attention for the first year or two, then you will reap the benefits later when plants are big and well-established enough to thrive without any extra care.

PLANTING IN HOT, SUNNY SITES

In some respects planting in the heat is a good thing, as a wilting plant is positively encouraged to take up water from the surrounding soil and may even get established faster. But a plant that is over-stressed may never fully recover, and this is when it can pay to look carefully and critically at the plant and cut off the largest leaves and flowering stems, which can make excessive demands on the root system and affect the whole plant.

If you are really worried, you can make a 'shade' from a frame of sticks covered with netting until the plant gets stronger or the weather cools down.

PLANTING IN DRY, SHADY AREAS

Dry shade is one of the most difficult sites to plant in, and even though you have chosen plants that have adapted to deal with this habitat, they will still respond to a bit of cosseting in the early stages of growth. The first step is to dig in some well-rotted manure or garden compost to improve the soil. Then water the soil well before and after planting. Try mulching the soil after heavy rainfall too, to keep moisture in. If you use an organic mulch, it will also help to improve the soil when it eventually rots down.

WATERING

Even drought-resistant plants need watering in their first year or two to help their roots get established. By giving these plants a helping hand while they are young, they will develop strong, healthy root systems that will subsequently deal with dry conditions all by themselves.

An irrigation system can make all the difference. It can be as sophisticated or as simple as you like, from an automated hose system connected to mains water and controlled by an app on your smartphone to a low-tech perforated hose connected to a rain barrel. In either case the hose can be discreetly set below the soil surface or covered with a mulch, both of which reduce evaporation.

If you're watering by hand, a full watering can of water once a week is better than a light sprinkling every day. Watering little and often, so that the water doesn't soak down to any great depth, encourages roots to stay near the surface, instead of spreading deep and wide in search of water.

PRUNING

Pruning not only shapes a plant to make it look its best, it also stimulates it into strong, healthy growth. Newly planted shrubs benefit from pruning. First of all snip out any dead stems. Then look for those that look spindly or weak, or where there are leaves showing signs of disease and stress. The final cuts you make should be to improve the plant's overall appearance and shape: prune the stems back to a fat, healthy bud that faces out, away from the plant. Plants with variegated leaves may produce shoots that have reverted to the original plain green form from which they have developed. These should be pruned out immediately, as they are often stronger than the variegated form and may weaken and take over the plant.

OPPOSITE Let cans of water stand for a couple of hours to reach ambient temperature before watering.

RIGHT Not all plants are pruned at the same time of year: check directory entry for specific advice.

HOW TO MAKE THE MOST OF DIFFICULT CONDITIONS

PLANTING IN AWKWARD SITES

In dry shade and where there are thirsty tree roots close to the soil surface, line planting holes with plastic to keep water close to the plant's roots rather than the tree's. Puncture the plastic in a few places so that excess water can drain away. If the soil is shallow or if there is a real mat of tree roots, try building up informal, small, raised beds for each plant. Make a mound of topsoil and well-rotted manure or compost on the surface of the ground, and keep it in position with a ring of rocks or stones.

PLANTING ON A SLOPE

Stabilizing the bare earth on a slope will give plants a better chance of establishing. Planting through plastic netting or a permeable membrane – held in place with metal pegs or skewers – will help. Cut a cross shape into the membrane where you want to add a plant and fold back the corners. Once the plant is in place, ease the corners underneath it.

Use a lawn sprinkler on its lowest setting to water a newly planted slope or you risk washing away precious topsoil.

MULCHING

A thick mulch helps to retain moisture around newly planted specimens as well as suppressing weeds, which compete for water, nutrients and light.

There is a wide range of organic mulches on the market – organic here meaning that they will eventually rot down – from chipped wood bark to leaf mold to spent mushroom compost,

LEFT Membranes, either permeable or impermeable, can help plants get established by stabilizing soil, suppressing weeds and retaining moisture. A gravel topping improves appearance.

ABOVE RIGHT A mulch of chipped bark will eventually biodegrade; gravel is longlasting but will still need topping up as it gets scattered.

even straw and seaweed. You can match the mulch to your site. To be effective, add a depth of at least 5 cm (2 in).

In hot, dry beds an inert gravel mulch is more in keeping and stops water evaporating. Gravel also warms up quickly in spring, giving sun-loving plants a boost. Pebble, slate and stone chippings have similar properties.

In winter a mulch gives plants an extra layer of insulation, and if it is deep enough can even prevent the soil from freezing, depending on how cold your winters are.

Apply mulch when the soil is wet after a good downpour. Never mulch frozen ground. And keep the mulch clear of plant stems or they could rot.

STAKING

In windy coastal gardens and exposed inland areas young trees and shrubs are going to need staking for a few years until their roots are strong enough to anchor them to the soil.

Choose a suitable size stake that holds the lower portion of the stem or trunk steady while letting the branches move freely in the wind. This encourages the stem to thicken and the roots to spread out to hold the plant in the soil.

Hammer the stake firmly into the planting hole before planting the tree or shrub to avoid damaging the roots.

Ideally, use a proper tree tie to hold the stem in place. If you prefer to improvise, use something soft, like cut-up strips of rag. Do not use wire or string, which can cut into bark and stems. Check ties twice a year, in autumn and spring, and loosen if necessary as the trunk grows thicker.

Perennials can often do with a bit of support on windy sites too, or even just to stop them bending and flopping after heavy rain. Again, it's best to get the support system into the ground at the same time as planting.

CONTROLLING AN OVERLY SUCCESSFUL PLANT

Sometimes a tough plant can be so successful that it becomes a nuisance. Various methods can be used to control it, depending on how it spreads. Plants that self-seed prolifically are relatively easy to deal with. Just keep an eye on them at flowering time and cut off the spent flower heads before they have time to set seed. Plants that spread by underground runners can be stopped in their tracks by a vertical layer of slates, tiles or bricks embedded in the border.

If you have a plant that behaves thuggishly, work with it. Use it as a signature plant and add it throughout the garden, rather than letting it spread unchecked in one border. But keep an eye on it.

WINTER PROTECTION

If some species in your garden are borderline hardy, don't leave their survival to chance. Use a thick layer of straw to protect crowns of perennials that die back to soil level or construct 'cages' from chicken wire around taller plants and stuff them with straw. Or wrap them with several layers of horticultural fleece, which lets air and light through but protects against extremes of temperature. Tie it in place with garden twine.

New plants that haven't made sufficient root growth can often be lifted out of the ground by a severe frost. When the soil freezes it expands, sometimes with enough force to lift a small plant along with it. After severe weather, walk around the garden and when the soil is no longer frozen, gently firm any affected plants back into place. If severe weather is forecast, prepare in advance and cover beds of vulnerable new plants with anything you have on hand: newspaper weighed down with bricks, old net curtains or sheets, or horticultural fleece.

Drought-resistant plants can be damaged by winter wet. Protect individual plants with plastic cloches pegged in place or improvise and make your own cloches or mini cold frames using old double-glazed doors or windows.

WARMING UP THE SOIL

If you want to get ahead with early planting in spring, pre-warming the soil gives plants a head start. Do this by covering flower beds with black plastic, clear plastic, cloches or horticultural fleece.

Depending where you live, you must judge when to cover the soil: it needs to

RIGHT Making the most of a hot dry site with pale stone and gravel to reflect the heat and sun and maximize conditions for Mediterranean plants such as lavender and thyme.

have been thoroughly soaked by winter rain but must not be frozen. Anchor plastic or fleece securely with pegs, bricks or by digging the edges into the soil or you risk high winds lifting it and wrapping it round a neighbor's tree.

CONSERVING WATER

Water is an increasingly precious resource that needs to be used wisely in the garden. Wherever you have a roof – whether it's the shed, conservatory or the house – install a rain barrel to collect rainwater. As well as saving water, the other big advantage is that water from a barrel is always at ambient temperature, so can be used straight away.

To reduce water loss by evaporation, water plants early in the morning or late evening. Established plants shouldn't need watering – save resources for newly planted beds and for trees and shrubs that you've planted within the past five years.

PLANT DIRECTORY

HOW TO USE THE DIRECTORY

The directory is divided into six habitats: sun, shade, dry, damp, exposed, coastal. In many cases the categories are not mutually exclusive and so the main entry for each plant is in the section where it tends to perform best, then it is cross referenced elsewhere.

The directory is by no means exhaustive. It is full of suggestions to inspire you and get you started. Some recommendations are for specific plants; others are just an example taken from a whole genus of tough plants that are worth growing.

A WORD ABOUT HARDINESS ZONES

This book uses the US Department of Agriculture system, which divides zones into 13 sections defined by the lowest minimum temperature in that region. The directory entries list the lowest temperature the plant will tolerate.

AVERAGE ANNUAL EXTREME MINIMUM TEMPERATURE

Temperature (F)	Zone	Temperature (C)
-60 to -50	1	-51.1 to -45.6
-50 to -40	2	-45.6 to -40
-40 to -30	3	-40 to -34.4
-30 to -20	4	-34.4 to -28.9
-20 to -10	5	-28.9 to -23.3
-10 to 0	6	-23.3 to -17.8
0 to 10	7	-17.8 to -12.2
10 to 20	8	-12.2 to -6.7
20 to 30	9	-6.7 to -1.1
30 to 40	10	-1.1 to 4.4
40 to 50	11	4.4 to 10
50 to 60	12	10 to 15.6
60 to 70	13	15.6 to 21.1

SITUATION

These symbols give a general indication of the problem situations that a plant will grow well in.

 Exposed Garden

 Sloping Garden

 Bog Garden

 Coastal Garden

MICROCLIMATE

These symbols indicate the extreme conditions a plant has evolved to cope with — some are able to cope with a range of these tough conditions.

 Moist, damp conditions

 Dry, drought conditions

 Sunny conditions

 Shady conditions

SOIL PH

If your soil has a high or low pH, choose plants that will flourish in particularly acidic or alkaline soil.

 Acidic soil

 Alkaline soil

TYPE OF SOIL

Once you have ascertained what type of soil you have, use these symbols to choose the plants that will thrive in your garden.

 Stony soil

 Clay soil

 Loam soil

 Sandy soil

 Chalky soil

 Poor soil

 Any soil

PREVIOUS PAGES A prairie-style planting of ornamental grasses, daisies such as helenium and other sun-loving perennials suits a hot, sunny garden with low rainfall.

OPPOSITE Hebes do well in coastal gardens, especially *Hebe albicans*, which forms a low, wind-resistant dome shape. It flowers in late spring and summer.

SUN

ACHILLEA FILIPENDULINA

Yarrow

ZONE
3-10

Yarrow is a tough wild flower of open fields and wastelands. *Achillea filipendulina* has finely cut, fern-like, soft green leaves. In summer it produces great flat heads of intense gold flowers on erect stalks, and carries on flowering well into autumn. Look out for cultivars such as 'Cloth of Gold' – at 1.8 m (nearly 6 ft) one of the tallest – and 'Gold Plate', just a little smaller and with grayer leaves.

- **HABITAT** sun, also suitable for dry, exposed, coastal
- **SOIL** sand, chalk
- **HEIGHT** 1.5 m (5 ft)
- **SPREAD** 1 m (3 ft)

WHERE TO PLANT

These plants will flourish in hot, dry sites and on sandy or thin chalky soils. The tough, wiry stems rarely need staking, even in exposed, windy sites or coastal gardens; they are flexible enough to bend and twist in the wind without snapping. Plant with other tough, summer-flowering perennials, such as poppies (*Papaver orientale*) and helenium hybrids, for a classic herbaceous border; plus some showy ice plants (*Sedum spectabile*) for late summer color.

CARING FOR PLANTS

To give them the best start, set out young plants in early spring. Divide older clumps when they start to get congested, flowering more densely around the edge than the center (usually after three or four years). Lift the clump, split it and replant the pieces separately, discarding any woody bits from the center. Do this straight after flowering or in early spring, just as the plant is coming into growth.

LEFT Make the most of a dry sunny garden with grasses and perennial sunflowers.

ALCHEMILLA MOLLIS
Lady's mantle

Its low-growing, clump-forming habit and the soft downy hairs on its leaves make lady's mantle a tough and adaptable plant. It's also pretty, with modest sprays of tiny, greenish-yellow flowers all summer long.

- **HABITAT**
  sun, also suitable for shade, exposed, coastal

- **SOIL**
  any

- **HEIGHT**
  50 cm (20 in)

- **SPREAD**
  50 cm (20 in)

WHERE TO PLANT

Lady's mantle can cope with an enormous range of situations. It is not fussy about soil – growing in sand, chalk or clay – as long as it is not waterlogged. It thrives in both hot sun and shade, brightening the shadows under trees and pushing up through unforgiving, sun-baked gravel. The low, rounded clumps resist the damaging effects of wind in exposed sites and coastal gardens.

CARING FOR PLANTS

Lady's mantle needs little attention, apart from dividing clumps when they get too big. Watch out for self-sown seedlings and, if necessary, move them to where you want them to grow. To avoid self-seeding, cut off the flower heads before the seed has a chance to ripen.

ALLIUM ATROPURPUREUM

Allium, ornamental onion

These bulbs belong to the same genus as onions, and you will notice this at once if you bruise a leaf. Most species like hot, dry conditions. *Allium atropurpureum* is one of the easiest to grow; it has heads of starry, purplish-red flowers in early summer, borne on erect stems to 1 m (3 ft) tall. Like most other alliums, it has long, strap-shaped leaves, which have a tendency to wither away before the flowers appear. Also look out for *A. cernuum*, which has loose, nodding heads of pink flowers, and *A. flavum*, which is lower growing and has yellow flowers. Pale lavender *A. hollandicum* will colonize a sunny bank.

- **HABITAT**
  sun, also suitable for dry, coastal

- **SOIL**
  any

- **HEIGHT**
  1 m (3 ft)

- **SPREAD**
  50 cm (20 in)

WHERE TO PLANT

A. atropurpureum will cope with dry, sun-baked earth and free-draining, sandy soils. Because the leaves tend to look untidy, mix it in with foliage plants that will disguise this habit. As long as the leaves have a reasonable amount of sun early in the growing season, it will not matter if they are shaded by neighboring plants later on.

CARING FOR PLANTS

Plant bulbs of *A. atropurpureum* in autumn. In general, plant any allium bulbs twice as deep as the width of the bulb.

Planting in autumn into an established border or bed helps you visualize more easily the impact the alliums will have when they are in full flower. Well-established clumps of allium bulbs can be divided in autumn, after flowering.

ANEMONE X HYBRIDA

Japanese anemone

See full entry on page 62

ANTHEMIS PUNCTATA
Sicilian chamomile

See full entry on page 87

ARTEMISIA
Wormwood

Wormwoods are best appreciated as foliage plants. The flowers are unprepossessing – small and a dull yellow. Their finely divided silver leaves are a foil for other plants in a mixed border while the smaller species make good ground cover. *Artemisia frigida* is low-growing, 8 cm (3 in), and forms a pretty silvery carpet of aromatic leaves. By contrast, the popular 'Powis Castle' is a dense bushy shrub that can grow up to 1 m (3 ft).

All artemisias are evergreen – or should that be eversilver?

- **HABITAT**
  sun, also suitable for dry

- **SOIL**
  Chalk, loam, sand

- **HEIGHT**
  1 m (3 ft)

- **SPREAD**
  1 m (3 ft)

WHERE TO PLANT

Set out young plants in spring in an open sunny spot – they're perfect for a gravel garden.

CARING FOR PLANTS

Prune plants in spring to keep them in good shape and to promote fresh new foliage. As many are not reliably hardy, take cuttings in summer so that you have a stock of replacements.

AUBRIETA DELTOIDEA
Aubretia

See full entry on page 88

BUDDLEJA DAVIDII
Butterfly bush

ZONE 5-10

This species of buddleja has long, arching stems and pointed, downy leaves, with white undersides. In summer it is smothered in graceful drooping spires of flower heads packed with tiny florets rich in nectar. One of the bonuses of the plant is that in addition to putting up with some awkward sites and conditions, it attracts masses of insects, particularly butterflies. Many cultivars have been developed from the original species, which has pale purple flowers. *B. davidii* 'Empire Blue' is purple-blue, 'Black Knight' a deep purple, and 'Royal Red' has reddish-purple flowers. Cultivars are usually smaller than the species, which can reach 3–4.5 m (10–15 ft).

WHERE TO PLANT
Butterfly bushes can be quick to colonize vacant ground. Take advantage of their tolerant, fast-growing nature and use them to create sheltered corners in an exposed garden or to make a patch of summer shade in an open space.

CARING FOR PLANTS
Butterfly bushes grow so fast that they can soon look messy if not maintained. Cut them back hard in spring, taking back every soft green shoot to within several centimeters (a few inches) of the hard, woody stems. This will ensure the best possible display of flowers and foliage.

An unseasonable frost may scorch young leaves and buds, but plants soon recover.

- **HABITAT**
  sun, also suitable for dry, exposed, coastal
- **SOIL**
  any
- **HEIGHT**
  4.5 m (15 ft)
- **SPREAD**
  2.5 m (8 ft)

CAMPANULA PERSICIFOLIA

Peach-leaved bellflower

Bell-shaped, purple flowers typify campanulas, while the plants themselves can be anything from low-growing, mat-forming species to *Campanula lactiflora* which is 1.5 m (5 ft) tall. In between extremes, *C. persicifolia*, the peach-leaved bellflower has slim, upright stems studded with blooms throughout summer 1 m (3 ft) tall. There are many cultivars, including 'Boule de Neige', with double white flowers, and 'Pride of Exmouth', which is pale blue.

- **HABITAT**
  sun, also suitable for shade

- **SOIL**
  any

- **HEIGHT**
  1 m (3 ft)

- **SPREAD**
  50 cm (20 in)

WHERE TO PLANT

Tough enough to put up with clay soils that crack in summer, this campanula is also tolerant of varying light, from full sun to medium shade. Its basal rosette of leaves can also help to stabilize a bare slope, paving the way for other plants once the soil has settled.

CARING FOR PLANTS

The flowering stems may need staking in open situations. If you cut them back after flowering you often get more flowers, but if you leave them to set seed you will get seedlings for next year.

CENTRANTHUS RUBER

Red valerian

See full entry on page 143

CERASTIUM TOMENTOSUM

Snow-in-summer

See full entry on page 89

CLEMATIS MONTANA
Clematis

Clematis montana is one of the toughest clematis. It flowers briefly and furiously in late spring and early summer, producing a great froth of white flowers. Cultivars bred from the species include 'Elizabeth' with vanilla-scented pink flowers; 'Mayleen' and *C. montana* var. *rubens* 'Odorata' are also pink and fragrant.

WHERE TO PLANT
Bleak bare walls, unsightly garages and ugly sheds all benefit from a curtain of clematis. All species prefer their roots to be kept cool with their heads in the sun. If growing against a wall set the plant at least 45 cm (18 in) away so that rain reaches its roots. *C. montana* can grow enormous with time – up to 4 m (13 ft) wide and twice as high – so site with care.

CARING FOR PLANTS
Spring-flowering clematis need little pruning but if *C. montana* does start to outgrow its situation, trim it lightly immediately after flowering. Mulch with well-rotted manure or garden compost in spring.

- **HABITAT**
  sun, also suitable for shade

- **SOIL**
  any

- **HEIGHT**
  12 m (40 ft)

- **SPREAD**
  4 m (13 ft)

ECHINACEA
Coneflower

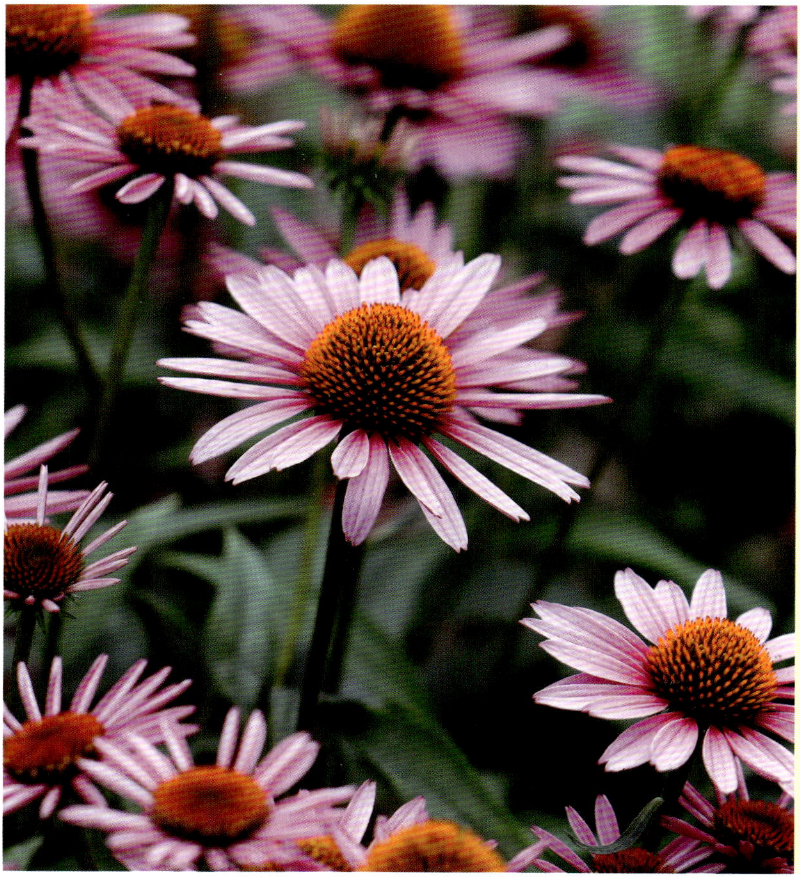

- **HABITAT**
  sun, also
  suitable for
  exposed, dry

- **SOIL**
  Any well
  drained

- **HEIGHT**
  1.5 m (5 ft)

- **SPREAD**
  50 cm (20 in)

Echinacea or coneflower is a North American prairie plant that has made its way into hot, sunny garden borders, often mixed with ornamental grasses and other daisy-type flowers such as gaillardia to recreate its natural setting. Its flowers are long-lasting and produced over many months. As well as the classic purple species, nurseries have bred varieties in shades of yellow, red, orange and white. All are great flowers for attracting pollinating insects.

WHERE TO PLANT

Echinaceas aren't fussy about soil and are tough sturdy plants that don't need staking. They can be up to 1.5 m (5 ft) tall so set them towards the back of a border or mix them in with other tall perennials. A spot in full sun is best: they will tolerate light dappled shade but won't flower so prolifically.

CARING FOR PLANTS

Start echinacea off from seed in spring under cover, planting out young plants after hardening them off for a few days. Mature plants form clumps that you can divide to make more plants if they start to take up too much space in the border. Do this in autumn or spring. Mulch established clumps with well-rotted manure or garden compost in autumn.

Deadheading encourages plants to produce more blooms – or pick yourself a bunch for the house as they make good cut flowers. Leave some flower heads to set seed for next year and also for the birds to eat.

ECHINOPS RITRO
Globe thistle

ENKIANTHUS CAMPANULATUS
Enkianthus

ERICA CINEREA
Heath, bell heather

ERYNGIUM BOURGATII
Sea holly

ESCALLONIA
Escallonia

EUPHORBIA

Spurge

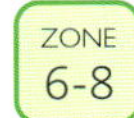

There's a euphorbia for almost every situation, from sun and shade to ground cover and subshrubs. They are not classical flowering plants and their flowers are technically bracts, a type of leaf, so are always shades of green or yellow-green. Magnificent *Euphorbia characias* subsp. *wulfenii* can reach 1.2 m (4 ft) and the flower heads alone can be 30 cm (1 ft) long.

- **HABITAT**
  sun, also suitable for dry shade
- **SOIL**
  Chalk, loam, sand
- **HEIGHT**
  1.2 m (4 ft)
- **SPREAD**
  1.2 m (4 ft)

WHERE TO PLANT

For a real burst of acid yellow-green in spring plant *E. characias* subsp. *wulfenii* in full sun. To continue the color theme into shadier areas use *E. amygdaloides*, a much scaled-down version just 60 cm (2 ft) tall.

CARING FOR PLANTS

When stems have finished flowering cut them down to ground level to make space for next year's shoots. Divide over-large clumps in spring or autumn. Wear gloves when handling euphorbias as the milky sap can irritate skin.

GAILLARDIA

Blanketflower

See full entry on page 132

HEBE ALBICANS

Hebe

See full entry on page 133

HELENIUM
Sneezeweed

ZONE
3-8

The rich coppery red flowers of *Helenium* 'Moerheim Beauty' will set a border ablaze. They are daisy-shaped with dark brown centers raised up into a central boss, set above a mass of mid-green leaves. 'Wyndley' has the same velvety dark brown centers to its flowers, but its yellow petals are splashed with orange.

- **HABITAT**
  sun
- **SOIL**
  any
- **HEIGHT**
  1 m (3 ft)
- **SPREAD**
  50 cm (20 in)

WHERE TO PLANT

These are plants of the North American prairies. They tolerate heavy clay without giving up and also do well on thin, chalky soil. Put them in full sun, and they will reward you with masses of flowers from early summer to early autumn. The average height is 1 m (3 ft), making them ideal mid-border plants.

CARING FOR PLANTS

When planting, put in some stakes or twiggy sticks to support the flowers. Regular deadheading will prolong the flowering season and boost the number of flowers produced. After two or three years clumps can become congested and will not flower so well. Divide and replant to revitalize them.

HIPPOPHAE RHAMNOIDES
Sea buckthorn

See full entry on page 148

HYPERICUM CALYCINUM
Rose of Sharon

See full entry on page 96

LAVANDULA ANGUSTIFOLIA
Lavender

Lavender deserves a place in every garden with a hot, sunny space to fill, and the late-summer flowers are a favorite with gardeners, bees and butterflies alike. The narrow, gray-green foliage and the fragrant, volatile oils it releases indicate that this is a plant well adapted to full sun and dry soil. Its dried flowers have been popular for centuries, and for the best drying results cut them before the buds are fully open.

WHERE TO PLANT
Try lavender as a low-growing hedge, clipping it to 60 cm (2 ft) tall, to separate areas of a garden or to edge a path or patio. Put it somewhere you visit regularly so that you can appreciate its fragrance as you pass. Like many aromatic Mediterranean plants, lavender must have full sun to flourish. It does not like winter wet and cold, so soil must be free-draining to avoid roots rotting.

CARING FOR PLANTS
Once the flowers are finished in early autumn, clip lightly all over with shears, taking off the flower stalks and a little foliage. Leave any serious pruning until spring, because the resulting new shoots are easily damaged by frost. When reshaping an old bush, never cut into the woody stem, because it is highly unlikely to regenerate. Ultimately it's better to dig out an old shrub that's past its prime and replant with a new one.

- **HABITAT**
  sun, also suitable for dry
- **SOIL**
  chalk, loam, sand
- **HEIGHT**
  1 m (3 ft)
- **SPREAD**
  1.2 m (4 ft)

Pictured right

SUN

MALUS × FLORIBUNDA

Japanese crab apple

ZONE 4-9

In spring the Japanese crab apple is a mass of deep pink buds that open to reveal pale pink or even white petals. The genus *Malus* includes the domestic apple tree and the Japanese crab apple produces small yellow edible 'apples' in autumn, but their size means that they are best left as a source of food for birds throughout winter.

- **HABITAT**
  sun, also suitable for partial shade

- **SOIL**
  any

- **HEIGHT**
  12 m (40 ft)

- **SPREAD**
  8 m (26 ft)

WHERE TO PLANT

Trees will tolerate chalk or clay, provided the ground is not permanently waterlogged. Although it is a small tree, ultimately the canopy can be as wide as the tree itself, so site it somewhere where it won't be constrained and you can appreciate its shape.

CARING FOR PLANTS

Shape the crown of the tree while it is small and the resulting mature tree should need little or no pruning, apart from removing any dead branches in winter. Although trees are generally trouble free, they can succumb to honey fungus or fireblight.

NEPETA × FAASSENII
Catmint

See full entry on page 98

PEROVSKIA ATRIPLICIFOLIA

Russian sage

This aromatic silvery subshrub – it doesn't grow quite big enough to be classified as a shrub – with spires of flowers deserves to be much more widely grown in hot, sunny gardens, cottage garden borders or courtyard plantings. The small, almost furry, blue flowers attract all manner of pollinating insects into the garden. Plants start flowering later in summer, in August, and carry on until late September.

- **HABITAT**
  sun, also suitable for exposed, dry, coastal

- **SOIL**
  any well drained

- **HEIGHT**
  1.2 m (4 ft)

- **SPREAD**
  1 m (3 ft)

WHERE TO PLANT

Grow perovskia in a hot, sunny border or a gravel garden. It's a tough resilient plant and its open airy shape goes well with other sun lovers such as eryngiums and echinaceas. Set plants where you'll be able to brush against them and appreciate their lavender-like scent. Although they grow tall, their open framework of stems doesn't obscure neighboring plants so it's fine to plant them at the front of the border or along the edge of a path.

CARING FOR PLANTS

You'll get the best flowering display if you cut perovskia right back to the ground just as you would a hardy fuchsia. Do this in autumn in milder regions; leave stems in place for extra protection until spring in colder areas. The silvery stems also look good in the garden in winter. Easily propagated from cuttings in spring or summer.

PERSICARIA BISTORTA
Bistort, snakeweed

See full entry on page 119

POTENTILLA FRUTICOSA
Cinquefoil

Potentillas are compact, deciduous shrubs with a neat, rounded shape. They have a long flowering period, from late spring to early autumn. *Potentilla fruticosa* 'Elizabeth' is one of the easiest cultivars to grow. It has simple, primrose-yellow flowers. The leaves are tiny and glossy green and are arranged in groups of five.

WHERE TO PLANT
Potentillas flourish in fairly poor soil, as long as it doesn't get too waterlogged in winter. They flower most strongly in full sun but can put up with a little shade – for example, from a nearby tree in the afternoon. A row of shrubs makes a decorative low hedge about 1 m (3 ft) high.

CARING FOR PLANTS
Trim bushes lightly just after flowering and before winter. At the same time, cut out any dead stems or weak spindly growth.

- **HABITAT**
  sun, also suitable for dry
- **SOIL**
  poor
- **HEIGHT**
  1 m (3 ft)
- **SPREAD**
  1 m (3 ft)

Pictured right

SUN

SANTOLINA CHAMAECYPARISSUS

Cotton lavender

ZONE 7-10

Cotton lavender is a low-growing, rounded bush with woolly white stems and small, silver-gray leaves. In summer it has button-shaped, yellow flowers. Gardeners who grow cotton lavender for its silver foliage often trim off the flower heads before they open, especially if the color would clash with a purple and pink border. The cultivar 'Lemon Queen' has gray-green rather than silver foliage and is taller and wider. 'Nana' is smaller and suitable for a rock garden.

- **HABITAT**
  sun, also suitable for dry

- **SOIL**
  Chalk, loam, sand

- **HEIGHT**
  50 cm (20 in)

- **SPREAD**
  1 m (3 ft)

WHERE TO PLANT

The silver foliage is a clue: cotton lavender thrives in hot, sunny sites, where the soil is free-draining or even positively dry. Use plants as a low dividing hedge, setting them out about 30 cm (1 ft) apart, between different plantings or alongside a path.

CARING FOR PLANTS

In autumn, clip off old flower heads and tidy up any over-long shoots. Cut older plants back hard each spring. Although cotton lavender is hardy, it's one of those plants that has a finite lifespan in terms of appearance. After five or six years, bushes begin to look woody and leggy and are best replaced.

SEDUM SPECTABILE

Ice plant, stonecrop

See full entry on page 102

SEMPERVIVUM TECTORUM

Common houseleek

See full entry on page 103

STACHYS BYZANTINA
Lamb's ears

     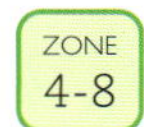

ZONE
4-8

Lamb's ears says it all: lovely, soft, strokeable woolly leaves in shades of gray. Even the summer flower spikes are woolly, studded with small pink flowers. The species is native to sun-baked soils in Turkey and Iran, and the furry silver leaves resist water loss and scorching.

WHERE TO PLANT
Use lamb's ears to underplant roses and disguise any less-than-lovely bare stems. Plant them as edging or in a mixed border. If you want ground cover without the flowers, plant the cultivar 'Silver Carpet'.

CARING FOR PLANTS
One of the few plants that can be said to be slug-proof – presumably they're not keen on furry leaves. If clumps get too large or overcrowded, divide them in spring or autumn.

- **HABITAT** sun, also suitable for dry
- **SOIL** chalk, loam, sand
- **HEIGHT** 50 cm (20 in)
- **SPREAD** 50 cm (20 in)

STIPA GIGANTEA
Giant feather grass, golden oats

See full entry on page 137

TAMARIX RAMOSISSIMA
Tamarisk

See full entry on page 153

TEUCRIUM CHAMAEDRYS
Germander

See full entry on page 104

THYMUS VULGARIS
Common thyme, garden thyme

See full entry on page 105

ULEX EUROPAEUS
Gorse, furze

See full entry on page 139

VERBASCUM BOMBICYFERUM
Mullein

See full entry on page 106

VERBENA BONARIENSIS
Argentinian verbena

ZONE 7-11

A trendy plant that has stayed the course and for good reason. Argentinian verbena is tall – up to 2 m (6½ ft) – and its tough wiry stems are topped with clusters of small, butterfly- and bee-friendly purple flowers. Its continued popularity has led plant breeders to develop smaller cultivars for smaller gardens: 'Little One' has the same airy open structure but to 40 cm (15¾ in); 'Lollipop' is the same height but the flowers stems branch low down.

WHERE TO PLANT
Use as an open screen to divide areas of the garden or to frame smaller flowering plants, or mix them with grasses and tall perennials in an informal prairie-style planting. Plant the smaller cultivars in open sunny flower beds.

CARING FOR PLANTS
Leave seed heads in place until spring – you can't have too much of *Verbena bonariensis*. Add a protective mulch of straw in winter in colder areas.

- **HABITAT** sun, also suitable for exposed
- **SOIL** any
- **HEIGHT** 2 m (6½ ft)
- **SPREAD** 50 cm (20 in)

Pictured left

VIBURNUM TINUS
Laurustinus, viburnum

See full entry on page 83

VINCA MAJOR
Greater periwinkle

See full entry on page 84

YUCCA FILAMENTOSA
Adam's needle

See full entry on page 155

SHADE

ANEMONE NEMOROSA

Wood anemone

ZONE
5-9

As its name suggests, the wood anemone is a woodland plant. It thrives in early season sun, before the tree canopy bursts into leaf, and thereafter enjoys the dappled shade of the leaf cover. It is low growing, with delicate, whitish, drooping flowers in spring. Wood anemones have deeply divided leaves and spread quickly to produce a carpet of ground cover.

Cultivars developed from the species include *Anemone nemorosa* 'Robinsonia', which has lavender flowers, and 'Blue Beauty', which has blue flowers and bronze-tinged foliage.

- **HABITAT**
  shade

- **SOIL**
  Chalk, loam, sand

- **HEIGHT**
  20 cm (8 in)

- **SPREAD**
  50 cm (20 in)

WHERE TO PLANT

Look for shady spots in the garden that resemble deciduous woodland – beneath shrubs or trees, for example, or in areas shaded by fences, sheds and other man-made structures. Take advantage of the anemone's quick-spreading nature to create a weed-free, groundcover carpet.

CARING FOR PLANTS

Plant rhizomes in autumn. If possible, dig in some well-rotted manure or some leaf mold to give them a good start. Add a mulch of leaf mold again in spring, especially if your anemones are growing in borders shaded by buildings, rather than under trees or shrubs, to mimic more closely woodland conditions. Autumn is also the best time to divide large clumps.

LEFT Woodland plants such as ferns and hostas thrive in dappled shade.

ANEMONE x HYBRIDA

Japanese anemone

Tough as old boots, say gardeners whose borders are overrun with Japanese anemones. To others they're a late-lowering treat. 'Honorine Jobert' is the classic white single flower; 'Königin Charlotte' has semi-double, frilly rose-pink blooms; and there are many more cultivars to choose from.

- **HABITAT**
  shade, also suitable for sun
- **SOIL**
  any
- **HEIGHT**
  1.2 m (4 ft)
- **SPREAD**
  1 m (3 ft)

WHERE TO PLANT

Shady or partially shaded borders, alongside north-facing walls or fences or anywhere that could do with a boost of color in autumn.

CARING FOR PLANTS

Water until established, in the first year or so. Dig out overspilling clumps, though you may not be able to transplant any surplus successfully as the roots are deep and brittle.

ASTILBE CHINENSIS

Astilbe

See full entry on page 109

AUCUBA JAPONICA

Spotted laurel

It is probably its adaptability that causes some people to sneer at spotted laurel. It can grow just about anywhere, particularly in neglected corners, and it may well appear miserable simply through its association with such less pleasant spots. It is, in fact, a very useful shrub that will grow to a rounded bush about 2.5 m (8 ft) tall. It can also be planted to create a dense and private hedge.

- **HABITAT** shade, also suitable for sun, dry
- **SOIL** chalk, loam, sand
- **HEIGHT** 2.5 m (8 ft)
- **SPREAD** 2.5 m (8 ft)

WHERE TO PLANT

If you must have a plant at all costs in a gray, sunless corner of the garden with soil so poor that even the weeds struggle, spotted laurel is for you. It will put up with just about anything, apart from permanently waterlogged soil, and its glossy evergreen leaves will reflect the light and look handsome all year. Cultivars of *Aucuba japonica* with variegated leaves, such as 'Crotonifolia' or 'Gold King', need some sun each day to keep their color or the leaves may eventually revert to plain green.

CARING FOR PLANTS

To get the characteristic, cheerful red berries in autumn you need a female plant with a male one in close proximity – either your own or one in a neighboring garden. If you spot any plain green leaves on variegated cultivars, cut the stems right back to the base to stop the plant from reverting.

CAMPANULA PERSICIFOLIA

Peach-leaved bellflower

See full entry on page 44

CAMELLIA

Camellia

Camellia x *williamsii* is one of the most easily grown camellias and one of the most reliable bloomers. Most of the cultivars have narrow, oval leaves and make tall, densely packed shrubs. The flowers vary in color from white to deep red and may be single or semi-double.

'Anticipation' has deep pink blooms reminiscent of peony flowers, with a central boss of stamens. It is recommended for containers and pots, making it suitable for patio gardens, although if grown directly in the soil it can get to 3 m (10 ft).

The weather determines precisely when in spring they flower, and a late winter can delay blooming by up to a month.

- **HABITAT**
  shade

- **SOIL**
  clay, loam,
  sand. Acidic

- **HEIGHT**
  4 m (13 ft)

- **SPREAD**
  4 m (13 ft)

WHERE TO PLANT

Many people go to great lengths to grow camellias in tubs and raised beds to create the acidic soil they prefer, so if you have acidic soil in your garden, you're one step ahead. The one proviso is to choose a spot in the garden that doesn't get the early morning sun. Although plants are reliably hardy, after a spring frost frozen flower buds can be damaged by thawing too quickly in strong sunlight.

CARING FOR PLANTS

Deadheading after flowering promotes strong growth, but no pruning is necessary other than to cut out spindly shoots or to reshape the odd branch. Mulching once or twice a year after a good downpour helps prevent the roots from drying out.

CHAENOMELES x SUPERBA

Japanese quince, flowering quince

This hybrid is a rather overlooked shrub, typical of suburban gardens. Yet it has a distinctly oriental pedigree; its parent plants are native to China and Japan. In spring it has cup-shaped blossoms, which come in various colors, ranging from orange and shading through red to pink and occasionally white. Named cultivars are more reliably consistent in color. The other common species, *Chaenomeles japonica* and *C. speciosa*, are equally useful in the garden. Japonicas may produce attractive apple-like fruits that ripen with a golden flush.

- **HABITAT**
  shade

- **SOIL**
  any

- **HEIGHT**
  1.5 m (5 ft)

- **SPREAD**
  2.5 m (8 ft)

WHERE TO PLANT

Japanese quince's lack of fussiness about soil types and light levels makes it a useful shrub. Although you might get more flowers in a sunnier spot, it still flowers reliably in the shade. Train shrubs on wires against a cheerless north-facing wall or incorporate them in a mixed hedge.

CARING FOR PLANTS

Prune wall-trained shrubs in spring by cutting back side branches to two or three buds. Otherwise, confine pruning to thinning out tangled branches or to keeping a shrub within bounds.

CHOISYA TERNATA

Mexican orange blossom

See full entry on page 90

CLEMATIS MONTANA

Clematis

See full entry on page 45

CORNUS ALBA
Dogwood

See full entry on page 111

COTONEASTER HORIZONTALIS
Herringbone cotoneaster

See full entry on page 91

CYCLAMEN HEDERIFOLIUM

Cyclamen

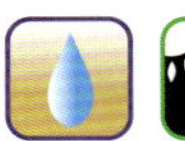   

ZONE 5-9

The flowers of this hardy cyclamen appear before the leaves and cover the ground with pale or deep pink, occasionally white, flowers in late summer and autumn. The flowers have elegantly reflexed petals, and when the seed pods ripen the flower stalks corkscrew towards the ground before bursting to scatter the seeds. The flowers are followed by deep green, heart-shaped leaves that last throughout the winter and fade at the end of spring.

- **HABITAT**
  shade, also suitable for dry

- **SOIL**
  any

- **HEIGHT**
  10 cm (4 in)

- **SPREAD**
  50 cm (20 in)

WHERE TO PLANT

Cyclamen do well in dry shade. Plant in autumn under shrubs or trees. The canopy of overhead leaves protects the tubers from excessive rain, which can cause rot. Plants rapidly build up into colonies by self-seeding and can form dense ground cover.

CARING FOR PLANTS

When the leaves wither in early summer, add a mulch of leaf mold. Cyclamen establish more readily if they are bought and planted as growing plants rather than dry tubers.

DAPHNE LAUREOLA

Spurge laurel

ZONE 6-9

In late winter and early spring spurge laurel has clusters of greenish-yellow flowers, half hidden in the axils of the leathery, glossy, dark green leaves. The flowers are sweetly scented and are followed by small black fruits.

- **HABITAT** shade, also suitable for dry
- **SOIL** any
- **HEIGHT** 1.2 m (4 ft)
- **SPREAD** 1.2 m (4 ft)

WHERE TO PLANT

Shade suits spurge laurel, and the shiny leaves reflect what little light there is in a gloomy corner. The soil can be anything from permanently damp to dry. The species makes a small, wide bush, about 1.2 m (4 ft) tall and up to 1.2 m (4 ft) across, and is useful for underplanting below taller shrubs or trees in a woodland setting.

CARING FOR PLANTS

Spurge laurels are not particularly long-lived shrubs, so grow some spares from cuttings taken in summer or keep an eye open for self-set seedlings. There's no need to prune, apart from taking out dead or spindly branches.

DRYOPTERIS FILIX-MAS

Male fern

The male fern is the least fussy of all ferns and tolerates a wide range of conditions. This makes it a versatile foliage plant, especially as it is semi-evergreen and looks good nearly all year round.

- **HABITAT** shade, also suitable for dry
- **SOIL** any
- **HEIGHT** 1.2 m (4 ft)
- **SPREAD** 1 m (3 ft)

WHERE TO PLANT

Use the male fern to bring life to dry, barren borders against a shady wall or to liven up the gloomy passage along the side of a typical townhouse. Mimic its natural woodland habit and plant it under trees or tall shrubs.

CARING FOR PLANTS

In poor conditions – full shade with dry soil – improve the soil with leaf mold or well-rotted garden compost and water well before planting. Make sure that the fern gets enough water over the next few months; then once it's established it will not need any further help. Cut back the old fronds once the new ones emerge so that you can appreciate their beauty as they unfurl.

ELAEAGNUS PUNGENS

Elaeagnus

See full entry on page 128

ENKIANTHUS CAMPANULATUS

Enkianthus

See full entry on page 113

EPIMEDIUM

Barrenwort

These delicate woodland plants with sprays of starry white, yellow or pink flowers are more resilient than they look. Toughest of all is *Epimedium* x *perralchicum* 'Fröhnleiten', with bright yellow flowers and truly evergreen leaves. *E.* x *versicolor* 'Sulphureum' has fresh green heart-shaped leaves that develop a red tinge as they age, turning bronze in autumn.

- **HABITAT**
  shade
- **SOIL**
  chalk, loam, sand
- **HEIGHT**
  50 cm (20 in)
- **SPREAD**
  1 m (3 ft)

WHERE TO PLANT

Plant in dappled shade under deciduous trees or shrubs or at the foot of a north-facing wall. Good for establishing on slopes and banks. Some species are more tolerant of dry soil than others.

CARING FOR PLANTS

Cut back last year's leathery leaves in spring to appreciate the flowers – though not if severe weather is forecast.

EUONYMUS FORTUNEI
Euonymus

See full entry on page 130

FILIPENDULA ULMARIA
Meadowsweet

See full entry on page 114

FUCHSIA MAGELLANICA
Magellan fuchsia

See full entry on page 131

GALIUM ODORATUM

Sweet woodruff

In late spring, sweet woodruff forms a carpet of starry, white flowers in the most unpromising situations. It is low growing, with leaves in a fresh shade of green, arranged around the stem in ruffs (technically whorls). A handful of stems makes a pretty posy, and it is such a prolific plant that cut stems often root in the vase. The flowers are scentless, but if the whole plant is dried, it develops the smell of new-mown hay.

- **HABITAT**
  shade, also suitable for dry

- **SOIL**
  any

- **HEIGHT**
  30 cm (1 ft)

- **SPREAD**
  1 m (3 ft)

WHERE TO PLANT

Sweet woodruff will grow just about anywhere, making it the perfect plant for awkward, dry, shady spots below shallow-rooted trees, such as silver birch and Norway maple, or for lightening the area at ground level under a mass of shrubs. It is an effective ground-cover plant – rather too effective, you may find, if it escapes into sunnier spots with richer soil. Divide clumps in autumn or spring if necessary.

CARING FOR PLANTS

Plants need minimal attention. Clip off old stems at the end of winter to neaten scruffy clumps, but be careful to avoid damaging emerging new shoots.

GERANIUM MACRORRHIZUM

Cranesbill geranium

See full entry on page 95

HEDERA HELIX

Common ivy

ZONE
3-10

Ivy barely needs any introduction. The glossy, evergreen leaves look good all year round, and although the flowers are hardly spectacular, they are an important source of nectar for bees and many other insects. The berries that follow are popular with birds and also provide winter color. *Hedera helix* is the species from which many cultivars have been developed, including 'Glacier', which has gray and cream variegated leaves, and yellow-edged 'Goldchild'. The cultivars are not as hardy.

- **HABITAT**
shade, also
suitable for dry

- **SOIL**
any

- **HEIGHT**
12 m (40 ft)

- **SPREAD**
4 m (13 ft)

WHERE TO PLANT

Plant ivy where it seems nothing else will grow. Use it as ground cover in the dense, dry shade cast by evergreen trees; squeeze it in where tree roots form an impenetrable layer just below the soil surface; or train it up a cold sunless wall.

CARING FOR PLANTS

Ivies are generally so vigorous that they can be cut back at any time of year if they start to overstep their limits. Sometimes the variegated varieties become less well marked in heavy shade, and soil that is too fertile can sometimes have the same effect.

"

HOSTA 'BIG DADDY'

Plantain lily

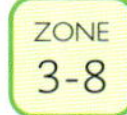

This hosta is a bit of a heavyweight, as its name suggests. It is one of the tougher, larger cultivars – to 1 m (3 ft) tall and 1 m (3 ft) across – and has leaves puckered like seersucker, with a true blue sheen. The flowers are pale, somewhere between violet and white, and are produced in summer.

- **HABITAT**
  shade
- **SOIL**
  clay, loam
- **HEIGHT**
  1 m (3 ft)
- **SPREAD**
  1 m (3 ft)

WHERE TO PLANT

Hostas prefer shade – bright sun can scorch their large, thin leaves – but it has to be moist shade. The density of the shade has a bearing on how prolifically they flower. In near-permanent shade you may have to appreciate them more for their beautiful foliage than for their flowers. They tolerate slightly acidic to neutral soils.

CARING FOR PLANTS

Before planting a hosta in heavy clay, work in a spadeful of gravel or grit in the bottom of the planting hole to help with drainage. Mulch plants in autumn with well-rotted manure or garden compost. Inspect plants regularly in the growing season – slugs and snails can devastate tender new shoots.

HYDRANGEA ANOMOLA *SUBSP.* PETIOLARIS

Climbing hydrangea

This climbing plant has the typical flower heads of a lacecap hydrangea. In the right circumstances, it may seem like the sky's the limit as it scrambles skywards – monsters of up to 12 m (40 ft) have been recorded. They can also spread the same distance horizontally. Flowering is most intense in the best light, so if your hydrangea climbs too high, you may miss out on its beautiful blooms at the top.

- **HABITAT**
  shade, also
  suitable for dry

- **SOIL**
  clay, loam

- **HEIGHT**
  12 m (40 ft)

- **SPREAD**
  8 m (26 ft)

WHERE TO PLANT

The climbing hydrangea is self-clinging and needs no extra support to scale walls, fences and even poles. It is ideal for gloomy alleys between houses, where its frothy white flowers lighten a shady wall in summer.

CARING FOR PLANTS

If you give this hydrangea enough space, you shouldn't need to prune it, apart from cutting out the odd awkwardly placed branch after flowering.

HYPERICUM CALYCINUM

Rose of Sharon

See full entry on page 96

LAMIUM MACULATUM
Deadnettle

ZONE 4-10

Lamium maculatum is closely related to its wild cousin *L. purpureum*. It has typical dead-nettle flowers with an upper and lower lip, toothed nettle-like leaves and square stems. It's odd how we regard one as a weed and the other as something to pay for. *L. maculatum* does have added extras, such as bigger flowers and attractive silver stripes on its leaves. Many cultivars have been developed, including 'Silver Nancy', which has leaves that are almost pure silver with a frill of green and white flowers, and 'Aureum', which has white-splashed gold leaves and pink flowers.

- **HABITAT**
  shade, also suitable for sun

- **SOIL**
  any

- **HEIGHT**
  20 cm (8 in)

- **SPREAD**
  1 m (3 ft)

WHERE TO PLANT

Use as ground-cover in both full shade and areas that receive some sun. It spreads quickly and makes a dense carpet of leaves. Lamium doesn't mind alkaline or chalky soil but it must be free-draining.

CARING FOR PLANTS

Cut back after flowering to encourage a fresh burst of weed-smothering growth. It spreads rampantly so keep it away from less vigorous species. Cultivars are slower growing.

LATHYRUS ROTUNDIFOLIUS

Persian everlasting pea

A tough and vigorous climber with small, sweet-pea-like flowers in an intense shade of purplish-pink, crammed on to long flowering stalks in summer. The leaves are dark green. More familiar is the everlasting pea (*Lathyrus latifolius*), which has slightly larger flowers, purple rather than pink, and which carries on blooming until well into autumn. Both species have curious 'winged' stems with a thin, leafy flange on each side.

- **HABITAT**
  shade

- **SOIL**
  any

- **HEIGHT**
  1.5 m (5 ft)

- **SPREAD**
  can spread widely

WHERE TO PLANT

The Persian everlasting pea prefers shade to sun. Grow it against a wall or fence with trellis or netting for support. The more common everlasting pea is slightly more versatile in that it will grow happily in sun or shade and can also be left to grow as ground cover or on banks and slopes, or even horizontally.

CARING FOR PLANTS

Pinch out the tips of plants in spring to encourage a thick, bushy shape. Keep deadheading flowers to prolong the display or cut them for the house for a short-lived bouquet – the color quickly fades in water. Cut plants right back to ground level in autumn if you're tidy-minded or wait until spring if not.

LONICERA SEMPERVIRENS

Trumpet honeysuckle, coral honeysuckle

One of the longest-flowering honeysuckle species, trumpet honeysuckle is in bloom from spring until autumn. Its clusters of striking red, tubular flowers, which give rise to its common name, gain extra emphasis from the leaves, which completely encircle the stem and the flower head in a chic, green frill. The technical term for this type of leaf is perfoliate. From late summer onwards, trumpet honeysuckle produces big, scarlet berries.

WHERE TO PLANT
Trumpet honeysuckle flowers better and more reliably in shade, so set the plant against a wall that is shaded for most of the day. Honeysuckles climb by twining and need some trellis tacked on to a wall or a system of horizontal wires to help them scramble up to cover bare bricks.

CARING FOR PLANTS
Trumpet honeysuckle is particularly susceptible to aphid attacks, although planting in shade helps to reduce this. It shouldn't really need pruning, but spindly, damaged or weak stems can be cut out in autumn. If your garden is in a particularly cold area, leave this until spring.

- **HABITAT**
  shade
- **SOIL**
  any
- **HEIGHT**
  4 m (13 ft)
- **SPREAD**
  1 m (3 ft)

Fictured left

MAHONIA AQUIFOLIUM
Oregon grape   See full entry on page 97

MALUS X FLORIBUNDA
Japanese crab apple   See full entry on page 52

PACHYSANDRA TERMINALIS

Pachysandra

An ideal ground-cover plant with oval, glossy, dark green leaves grouped together at the tops of its stems. In summer it bears small spikes of fragrant white flowers, which are sometimes followed by white berries. There is also a variegated variety, 'Variegata', with off-white edging to its leaves.

- **HABITAT**
  shade, also suitable for dry, damp

- **SOIL**
  any

- **HEIGHT**
  50 cm (20 in)

- **SPREAD**
  1 m (3 ft)

WHERE TO PLANT

Pachysandra never grows taller than around 50 cm (20 in). Its height, combined with its free-spreading habit – the stems root at intervals as they spread across the soil – make it good for filling bare ground in shade. The denser the shade, the more slowly it grows, but it will get there in the end. Don't let it spread into areas of good soil or you may find yourself with an invader on your hands.

CARING FOR PLANTS

Work in some leaf mold before setting plants out in early spring, especially if the soil is dry. Try not to let plants dry out in their first year or two, but thereafter they should cope with all but the worst conditions. Its habit of rooting at intervals makes cuttings easy to take and you can quickly increase your stock of plants.

PERSICARIA BISTORTA

Bistort, snakeweed

See full entry on page 119

PULMONARIA OFFICINALIS

Lungwort

ZONE
3-8

Lungwort has bristly leaves spotted with silvery white. In spring it produces sprays of small, drooping flowers that start off pink and fade to blue, giving a pretty two-color effect. Several cultivars have been developed, including 'Sissinghurst White' and 'Blue Mist'.

- **HABITAT**
  shade

- **SOIL**
  chalk, clay, loam

- **HEIGHT**
  30 cm (1 ft)

- **SPREAD**
  50 cm (20 in)

WHERE TO PLANT

Shade-loving pulmonarias make pretty ground cover below shrubs and trees. Some cultivars, such as 'Sissinghurst White', also tolerate full sun, but they will not do as well as in shade. As they are at their best in spring, do not give them too prominent a position in the garden because plants can look rather tired by midsummer.

CARING FOR PLANTS

Pulmonarias are prone to powdery mildew in dry weather. If plants become affected, cut the leaves right back to ground level and water well to encourage fresh new growth. Plants often self-seed if you don't cut them back after flowering.

RHEUM PALMATUM
Chinese rhubarb

See full entry on page 121

SHADE

RHODODENDRON 'LODERI GROUP'

Rhododendron

Rhododendrons are stunning spring shrubs and those in the 'Loderi Group' have large fragrant flowers in trusses of ten blooms or more. Many of its cultivars – 'Pink Topaz' and 'Venus' – can reach 4 m (12 ft).

- **HABITAT**
  shade
- **SOIL**
  clay, loam, sand. Acidic
- **HEIGHT**
  4 m (13 ft)
- **SPREAD**
  4 m (13 ft)

WHERE TO PLANT

Shade doesn't bother rhododendrons, which are largely woodland plants. What they must have without fail is acidic soil. They make stunning hedges and useful windbreaks.

CARING FOR PLANTS

Deadhead shrubs to prolong flowering and improve appearance. In soil that veers towards neutral pH, add an annual mulch of pine needles to increase the acidity. Improve the drainage of acidic clay soils by digging in sand and well-rotted garden compost or leaf mold before planting.

RODGERSIA PINNATA

Rodgersia

See full entry on page 122

RUSCUS ACULEATUS

Butcher's broom

See full entry on page 101

SARCOCOCCA CONFUSA

Sweet box

Known for its fragrance at a bleak time of year, sweet box, sometimes called Christmas box, is a dense evergreen bush. It flowers from midwinter on, the clusters of small, creamy flowers contrasting with the glossy, black fruits from the previous year – if the birds haven't already eaten them.

- **HABITAT**
  shade, also
  suitable for dry

- **SOIL**
  any

- **HEIGHT**
  1 m (3 ft)

- **SPREAD**
  1.5 m (5 ft)

WHERE TO PLANT

Sweet box puts up with deep, dry shade. To make the most of its scent, put it where you will not have to walk far to find it on cold days. Plants eventually grow to around 1 m (3 ft) tall and 1.5 m (5 ft) across.

CARING FOR PLANTS

For best results, plants need shelter from the coldest winds; siting them near a wall or below taller shrubs and trees will help. Pruning shouldn't be necessary, except to reshape or to remove dead wood, which should be done in spring.

SKIMMIA JAPONICA

Japanese skimmia

See full entry on page 152

SYMPHYTUM

Comfrey

Native wild plants, comfreys lend themselves to wildlife areas and neglected corners of the garden. Common comfrey, *Symphytum officinale*, is bristly with purple flowers. *S. grandiflorum* really earns the title of tough plant. Low-growing, with oval green leaves and sprays of creamy white flowers in spring, it spreads to make ground cover in unpromising dry shade and has a reputation for holding its own against ground elder.

- **HABITAT**
  shade
- **SOIL**
  any
- **HEIGHT**
  1 m (3 ft)
- **SPREAD**
  1.5 m (5 ft)

WHERE TO PLANT

Grow *S. grandiflorum* under shrub or tree canopies or in shady unloved borders down the side of townhouses.

CARING FOR PLANTS

Give new plants a boost with leaf mold and water until established, then maintenance free. May become too rampant in some gardens.

VIBURNUM TINUS

Laurustinus, viburnum

Whatever the time of year, there's always something to appreciate with laurustinus. The pink flower buds form from autumn onwards and open over a long season from late winter to spring. The flower heads are flat clusters of pinkish-white flowers. There are often flowers and blue-black berries on the bush at the same time. The evergreen leaves are dark green and shiny.

Viburnum x burkwoodii is another hardy evergreen that has scented flowers and downy brown undersides to its leaves. Other useful species include V. sargentii, which has red autumn berries; V. plicatum with its big, flat flower heads that are rather like hydrangeas; and V. farreri, which has new bronze foliage turning to bright green as it matures. All three are deciduous.

- **HABITAT**
  shade, also suitable for sun, dry, coastal

- **SOIL**
  any

- **HEIGHT**
  2 m (6½ ft)

- **SPREAD**
  2 m (6½ ft)

WHERE TO PLANT

Try growing any of these species as a hedge, clipped formally or left to spread. They are also good-looking enough to be a feature in a shrub border or by a garden entrance. None of these species minds shade.

CARING FOR PLANTS

Trim hedges and prune formal hedges severely into shape after the main flowering in late spring. For specimen bushes, cut some of the older stems back to ground level to encourage new growth.

VINCA MAJOR

Greater periwinkle

Periwinkle has long, questing shoots which rapidly colonize bare earth. 'Variegata' has dark green, oval leaves edged with creamy white. The flowers are violet and are produced intermittently throughout the year. It will scramble through taller plants or over low walls and fences. Periwinkles with different color flowers mainly belong to the closely related species, lesser periwinkle, *Vinca minor. V. minor* f. *alba* has white flowers; 'Atropurpurea' reddish-purple.

- **HABITAT** shade, also suitable for sun

- **SOIL** any

- **HEIGHT** 30 cm (1 ft)

- **SPREAD** 2.5 m (8 ft)

WHERE TO PLANT

Periwinkle is useful for quickly covering a bare, shady bank, but plants may overstep their limits, so take care not to let them spread out of control. The denser the shade, the fewer flowers you will get, but variegated leaves add interest. Periwinkle tolerates thin, chalky soils.

CARING FOR PLANTS

Cut plants back hard in spring to keep them under control. If you do not want periwinkle to spread any further, keep pulling back long shoots, which will root wherever they touch the ground.

RIGHT Periwinkle is a plant for difficult dry shade such as that under mature trees and for covering sloping sites.

SHADE

DRY

ANTHEMIS PUNCTATA *SUBSP.* CUPIANANA
Sicilian chamomile

 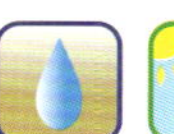    

ZONE 4-9

A fairly low plant that forms a dense evergreen mat. It has pleasantly aromatic, silver-gray foliage and in summer blooms with a great flush of daisy-like, white flowers with yellow centers that last and last. In a temperate winter the foliage dulls to a gray-green but does not die away.

- **HABITAT**
  dry, also suitable for sun, coastal

- **SOIL**
  chalk, loam, sand

- **HEIGHT**
  50 cm (20 in)

- **SPREAD**
  1 m (3 ft)

WHERE TO PLANT

Because it never grows higher than about 50 cm (20 in), anthemis is best planted at the edge of a border or along a path. Set out young plants in spring in a hot, sunny position. The plants do not mind chalky or sandy soil. Their mat-forming habit and silver summer foliage make them ideal in coastal gardens, although they do prefer a spot in the shelter of larger plants if possible. They also do well in containers.

CARING FOR PLANTS

To keep anthemis plants growing vigorously, cut them back hard after flowering or, at the very least, snip off the flowering stems. If you're worried about plants overwintering, take cuttings of young shoots in spring and grow them on in pots for replacements.

Divide plants in spring or autumn if they outgrow their space.

LEFT Succulents such as sedums and sempervivums have fleshy leaves that can store water and their waxy surface prevents them drying out.

AUBRIETA DELTOIDEA

Aubrieta

     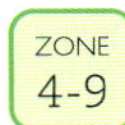

In some areas, cascades of intense purple or cerise flowers on old garden walls are a sure sign of spring. Aubrieta plants form great, ground-hugging mats of color. All hybrids come from the original species, *Aubrieta deltoidea*, which is itself quite variable. The flower colors range from white to violet, and some are tinged with red. They are simple, four-petalled flowers, but massed together in a mature plant they create a huge impact.

- **HABITAT**
  dry, also suitable for sun

- **SOIL**
  chalk, sand

- **HEIGHT**
  10 cm (4 in)

- **SPREAD**
  1 m (3 ft)

WHERE TO PLANT

Aubrietas come from dry, rocky, mountain slopes in the Middle East and are perfectly at home rooting in tiny cracks in paving or gaps where mortar has fallen from a low wall. They tend to be too vigorous for a rockery full of alpines and are better used to soften hard-edged landscaping. But don't plant them directly on a paved path as they won't survive being walked on.

CARING FOR PLANTS

Clipping mats of aubrieta with garden shears after flowering promotes healthy growth and stops plants from becoming too straggly. Aubrieta will often self-seed; transplant seedlings before they have time to wedge themselves into inaccessible crevices.

AUCUBA JAPONICA

Spotted laurel

See full entry on page 63

BRACHYGLOTTIS MONROI

Monro's ragwort

See full entry on page 141

BUDDLEJA DAVIDII
Butterfly bush

See full entry on page 43

BUPLEURUM FRUTICOSUM
Shrubby hare's ear

See full entry on page 142

CERASTIUM TOMENTOSUM

Snow-in-summer

ZONE 2-8

A vigorous mat-forming perennial with narrow leaves and stems. Both are so densely furred with hairs that they look white from a distance. In late spring and summer it has masses of white, starry flowers. Rarely more than 30 cm (12 in) tall.

- **HABITAT**
  dry, also suitable for sun
- **SOIL**
  any
- **HEIGHT**
  30 cm (12 in)
- **SPREAD**
  1 m (3 ft)

WHERE TO PLANT

Its rampant spreading habit makes it ideal for stabilizing and disguising a bank of dry soil or for covering up a less-than-beautiful wall. On level ground its dense growth makes it an effective weed-suppressing mat, especially as the leaves are evergreen, but keep it away from less dynamic species, which may get swamped.

CARING FOR PLANTS

Plants need little attention, but clipping lightly after flowering will keep them in good shape. Control them by digging up and dividing as necessary, either in spring or late in summer.

CHOISYA TERNATA

Mexican orange blossom

This bushy shrub grows as wide as it is high, and established plants can get to 2 m (6½ ft) tall. The glossy, evergreen leaves have a pleasant scent when crushed, and in spring the clusters of small white flowers have a delicious citrus perfume.

- **HABITAT**
  dry, also suitable for sun, shade and coastal

- **SOIL**
  any

- **HEIGHT**
  2 m (6½ ft)

- **SPREAD**
  2 m (6½ ft)

WHERE TO PLANT

It is unfussy about soil type, being able to cope with thin, chalky soils or clay that bakes hard in summer. 'Sundance', the golden-leaved cultivar of *Choisya ternata*, really does need full sun to keep its vibrant gold color, but the ordinary species can be grown in a shady corner. Plant choisya as a shrub or use it as part of a mixed hedge.

CARING FOR PLANTS

Plant new shrubs in spring so that they have all summer and autumn to get established and firmly rooted before winter. On exposed sites, frosts can scorch new growth, but plants generally recover. There is no need to prune unless the plant oversteps its boundaries.

CISTUS X CYPRIUS

Rock rose, sun rose

See full entry on page 144

COTONEASTER HORIZONTALIS
Herringbone cotoneaster

This cotoneaster is an unassuming little shrub, tenacious in its ability to grow in inhospitable sites. The genus also contains some evergreen species. *Cotoneaster dammeri* grows in similar habitats to *C. horizontalis* and is similarly useful but low-growing, to 8–10 cm (3–4 in) maximum.

- **HABITAT**
  dry, also suitable for sun, shade and exposed

- **SOIL**
  any

- **HEIGHT**
  1 m (3 ft)

- **SPREAD**
  1.5 m (5 ft)

WHERE TO PLANT
As its name suggests, *C. horizontalis* spreads wider than it grows upwards. It will probably not grow more than 1 m (3 ft) high but may spread to 1.5 m (5 ft). It can grow well in dry shade, on bare slopes and against cold north walls. It often self-seeds in unlikely places, saving you the job of deciding where to plant it.

CARING FOR PLANTS
There shouldn't be any need to prune, but if you need to take out any awkward branches do this in late winter. Get new plants into the ground in spring to help them become established before winter.

CRAMBE MARITIMA
Sea kale

See full entry on page 145

CYCLAMEN HEDERIFOLIUM
Cyclamen

See full entry on page 66

CYTISUS SCOPARIUS
Common broom

See full entry on page 146

DAPHNE LAUREOLA
Spurge laurel

See full entry on page 67

DRYOPTERIS FILIX-MAS
Male fern

See full entry on page 68

ECHINACEA
Coneflower

See full entry on pages 46–7

ECHINOPS RITRO

Globe thistle

      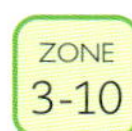

Globe thistles are typical drought-resistant plants. They have narrowly divided, spiny leaves with silvery undersides that help to reduce moisture loss. The stems are silver-gray, too, and contrast with the steel-blue flower heads, which open in succession from midsummer to autumn. Globe thistles make good cut flowers and can be dried for winter arrangements.

- **HABITAT** dry, also suitable for sun

- **SOIL** chalk, loam, sand

- **HEIGHT** 1.5 m (5 ft)

- **SPREAD** 1 m (3 ft)

WHERE TO PLANT

Globe thistles must have full sun to thrive and develop. Use them to stabilize a dry, sandy bank, where the basal rosette of leaves will help to resist soil erosion, or plant them to create shelter for smaller plants in an open, exposed garden.

CARING FOR PLANTS

All plants that are grown in tough conditions benefit from a little initial cosseting until they become established, and globe thistles are no exception. Once they have put down strong roots, however, there will be no stopping them. Globe thistles tolerate poor soils, so there is no need to add any well-rotted manure or garden compost, but if the soil is heavy, work in some gravel before planting. Keep plants well watered in their first season. Globe thistles soon form large clumps, which can be divided to give even more coverage.
   Aphids can be a nuisance.

Pictured right

ELAEAGNUS PUNGENS
Elaeagnus

See full entry on page 128

ERICA CINEREA
Heath, bell heather

See full entry on pages 128–9

DRY

ERYNGIUM BOURGATII
Sea holly

See full entry on page 147

EUONYMUS FORTUNEI
Euonymus

See full entry on page 130

EUPHORBIA
Spurge

See full entry on page 48

GAILLARDIA
Blanketflower

See full entry on page 132

GALIUM ODORATUM
Sweet woodruff

See full entry on page 70

GERANIUM MACRORRHIZUM
Cranesbill geranium

ZONE 4-9

These geraniums are not the familiar tender container plants, but are hardy perennials that do well in sun or shade. Soft, downy leaves and purplish, pale pink or white flowers characterize this species. It flowers in spring, and when the petals drop the developing seed heads still have a certain charm. They have a bulbous base and a long pointed 'beak', hence the common name. The leaves are strong smelling, having a vaguely antiseptic scent mixed with a hint of cat urine, but surprisingly not entirely unpleasant.

WHERE TO PLANT

Geranium macrorrhizum thrives in poor, dry soil in shade, vigorously growing to form an attractive clump. Plants also self-seed to make a spreading ground-cover colony. In heavier shade, the much paler flowers of the cultivar 'Album' will lighten the gloom. In warmer areas, plants keep their leaves all year round and the old foliage often turns color in autumn, typically to rusty red.

CARING FOR PLANTS

Cut off the seed heads if you don't want more seedlings. Trim back any bare, trailing stems to promote new growth, but generally these plants are trouble-free.

- **HABITAT**
  dry, also suitable for shade
- **SOIL**
  any
- **HEIGHT**
  30 cm (12 in)
- **SPREAD**
  1 m (3 ft)

Pictured left

HEDERA HELIX
Common ivy

See full entry on page 71

HYDRANGEA ANOMOLA SUBSP. PETIOLARIS
Climbing hydrangea

See full entry on page 73

HYPERICUM CALYCINUM
Rose of Sharon

The bright yellow, cup-shaped flowers of this evergreen shrub are borne for a long period, starting at the beginning of summer and carrying on well into autumn. There are other species of hypericum, but *Hypericum calycinum* is one of the hardiest, along with *H. androsaemum*, which has smaller flowers and is deciduous.

- **HABITAT** dry, also suitable for shade
- **SOIL** any
- **HEIGHT** 30 cm (1 ft)
- **SPREAD** 1.5 m (5 ft)

WHERE TO PLANT

Rose of Sharon is a useful ground-cover species. Plant it to form a lowish carpet, with a maximum height of 30 cm (1 ft), below taller shrubs or under trees in a woodland setting. You can grow it as groundcover in the sun, too, where it will flower more prolifically.

CARING FOR PLANTS

Deadheading prolongs flowering. For vigorous, bushy growth, trim back plants in spring to create a neat shape. Older specimens benefit from being cut back hard to the ground at the same time of year, which will rejuvenate them. In a particularly cold winter, plants of *H. calycinum* may lose their leaves.

MAHONIA AQUIFOLIUM

Oregon grape

Oregon grape has glossy, dark green leaves with spiny, serrated margins, rather like a holly leaf. As winter sets in, the leaves turn a subtle shade of purplish-brown. At the same time, tantalizing clusters of flower buds appear, but the shrub doesn't properly burst into bloom until late spring. When the flowers finally open, you will certainly know, as they have a delicate lily-of-the-valley scent.

WHERE TO PLANT

The Oregon grape is a versatile shrub which can cope with a surprising range of conditions, from clay to sandy or chalky soil and from sun to shade, without showing any effect on flowers, foliage or growth. The species *Mahonia aquifolium* forms a bush eventually 1.5 m (5 ft) tall, but there are several lower-growing cultivars that make good ground-cover plants. 'Apollo' reaches no more than 60 cm (2 ft) tall and often produces black berries after flowering. 'Smaragd' is similar.

- **HABITAT**
  dry, also suitable for sun and shade

- **SOIL**
  any

- **HEIGHT**
  1.5 m (5 ft)

- **SPREAD**
  1.5 m (5 ft)

CARING FOR PLANTS

Grow a mahonia as a feature shrub, and it will not need pruning, apart from removing any straggly growth in either autumn or spring. It is prone to suckering, so if you want to keep a neat shape, the suckers can be dug up and replanted elsewhere or simply pulled out and composted. If you are using low-growing cultivars as ground cover, shear them off at ground level every two or three years. Do this after they have flowered.

NEPETA x FAASSENII

Catmint

     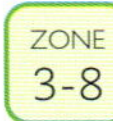

The wrinkled, greenish-gray leaves of catmint are covered with soft hairs – a defense against drought – and are also strongly aromatic. Cats love the smell. Its pale purplish-blue flowers are arranged in spikes to 50 cm (20 in) tall. The cultivar 'Six Hill's Giant' has deeper purple flowers and grows to 1 m (3 ft) tall.

- **HABITAT**
  dry, also suitable for sun, exposed and coastal

- **SOIL**
  any

- **HEIGHT**
  50 cm (20 in)

- **SPREAD**
  50 cm (20 in)

WHERE TO PLANT

Catmint is a deceptively pretty plant that is also a tough customer. It likes hot, dry sites and free-draining sandy or chalky soils, but will also grow in clay. It forms dense clumps that make weed-suppressing ground cover.

CARING FOR PLANTS

Cutting back flowering stems once they're finished often stimulates another flush of blooms. Clip the whole plant lightly at the same time if growing it for ground cover. In spring cut out any old woody stems to rejuvenate plants.

ONOPORDUM ACANTHIUM

Scotch thistle, cotton thistle

ZONE
6-10

A colossal ornamental Scotch thistle in full flower and leaf is an arresting sight. Getting close to 2.5 m (8 ft) tall and around 1 m (3 ft) across, it produces typical thistle-type flower heads, which are quite modest in size by comparison to the plant's overall size. Everything about the plant is spiny, from its silvery leaves to its 'winged' stems.

- **HABITAT** dry, also suitable for exposed
- **SOIL** chalk, loam, sand
- **HEIGHT** 2.5 m (8 ft)
- **SPREAD** 1 m (3 ft)

WHERE TO PLANT

Scotch thistle is at home in dry gravel beds and hot, sunny borders. For obvious reasons, give it plenty of space and do not site it close to a path – a spiny embrace from a thistle is something you can do without. Similarly, Scotch thistles and small children do not go together.

CARING FOR PLANTS

Scotch thistles are biennials, so for the first year of life they simply form a flat rosette of prickly leaves. Set plants out at the end of summer or beginning of autumn for flowering the following summer. Cut off most of the seed heads to avoid a forest of seedlings springing up, but leave a few if you want more plants for subsequent years. Alternatively, try cutting the flowers for drying.

PIERIS 'FOREST FLAME'
Pieris

One of the main charms of pieris is its rich red new foliage in spring, which gradually fades to pink and finally a more prosaic green. When it flowers it has pretty sprays of white, bell-shaped blooms, but it does not always do this reliably. If you want flowers, *Pieris japonica*, the lily-of-the-valley bush, is the one you need, but it is not as reliably hardy in colder areas as 'Forest Flame'.

- **HABITAT**
  dry, also
  suitable for
  sun

- **SOIL**
  loam, sand.
  Acidic

- **HEIGHT**
  3 m (10 ft)

- **SPREAD**
  2.5 m (8 ft)

WHERE TO PLANT

All species and cultivars of pieris must have acidic soil if they are to thrive. 'Forest Flame' will grow in full sun or in part shade and will tolerate a certain degree of dry soil – in a woodland area or below taller shrubs, for example. 'Forest Flame' can grow quite tall itself if you let it, sometimes topping 3 m (10 ft) after many years.

CARING FOR PLANTS

It's not unusual for the first flush of new leaves to be stricken by a late spring frost, but plants should recover. Wait until summer to cut back frost-damaged shoots. If you need to prune to reshape the plant, do this right after the shrub has flowered or in early summer.

RUSCUS ACULEATUS

Butcher's broom

The charms of butcher's broom are rather understated. It has tough, green leaves, which are actually flattened stems. In spring tiny, star-shaped, green flowers appear to sit on the surface of the 'leaf'. Male and female flowers are borne on separate plants, and if you grow both, the female plants produce scarlet berries. The real virtue of butcher's broom is that it grows in the most inhospitable dry soil. Even if you haven't seen it growing before, you have probably come across it as a foliage plant in florist's bouquets.

- **HABITAT**
  dry, also
  suitable for
  sun and shade

- **SOIL**
  any

- **HEIGHT**
  1 m (3 ft)

- **SPREAD**
  1 m (3 ft)

WHERE TO PLANT

Grow butcher's broom in those shady 'black spots', such as beneath evergreen trees, shrubs or shallow-rooted trees such as silver birch. Space plants out so that they are about 60 cm (2 ft) apart and they'll make good ground cover.

CARING FOR PLANTS

Plant a mix of male and female plants if you want a show of berries in autumn. Cut back any dead or straggly stems in spring.

SEDUM SPECTABILE

Ice plant, stonecrop

This late-flowering stonecrop brings a burst of color to the garden in autumn. The large, flat flower heads are packed with tiny pink flowers. As soon as they open, they attract dozens of butterflies and bees. Even after the flowers have long faded, the dried stems and heads still look interesting, especially when etched with frost. The leaves are gray-greenish and thick and fleshy, and the whole plant overwinters as a densely packed mass of nubby buds. There are several cultivars, which differ mainly in flower color: 'Carmen' has crimson flowers, 'Brilliant' is a deep rose pink, and 'Iceberg' is white.

- **HABITAT** dry, also suitable for sun
- **SOIL** chalk, loam, sand
- **HEIGHT** 60 cm (2 ft)
- **SPREAD** 60 cm (2 ft)

WHERE TO PLANT

The thick, fleshy leaves of stonecrop indicate its capacity to conserve water in hot, dry conditions. Set it in full sun in free-draining soil. Waterlogged winter soils can cause the roots and the whole plant to rot. Individual clumps seldom exceed 60 cm (2 ft) in either height or width.

CARING FOR PLANTS

Leave the dead flower stems and leaves in place over winter to give a bit of extra protection to the developing buds below. In spring, old stems should pull away easily, without any resistance.

SEMPERVIVUM TECTORUM

Common houseleek

ZONE
2-11

The bristly tipped leaves of the common houseleek are arranged in neat overlapping circles to form a rosette. Mainly blue-green, the leaves sometimes become flushed with color in summer and turn a subtle shade of red. At the same time, plants send up flowering stems topped with clusters of starry red flowers. The rosettes are low growing, typically 15 cm (6 in) high, but the flowering stems can reach 50 cm (20 in).

- **HABITAT**
dry, also
suitable for
sun

- **SOIL**
loam, sand

- **HEIGHT**
15 cm (6 in)

- **SPREAD**
50 cm (20 in)

WHERE TO PLANT

Houseleeks can sometimes be found growing on old tiled roofs and garden walls. Their succulent leaves conserve water, and the way they overlap in a rosette also helps the leaves shade each other from the sun. Plant in full sun in the driest possible spot. They will rot in cold, wet winter soil.

CARING FOR PLANTS

An individual rosette that has flowered will die, but there should be plenty of offsets to make a display for years to come. Look out, too, for *Sempervivum ciliosum*, which has hairy leaves and yellow flowers and enjoys the same conditions.

SKIMMIA JAPONICA

Japanese skimmia

See full entry on page 152

STACHYS BYZANTINA

Lamb's ears

See full entry on page 57

TEUCRIUM CHAMAEDRYS

Germander

ZONE 5-11

Germander has aromatic foliage, and its shiny, green leaves are thickly felted below with gray hairs. In summer it produces spikes of red, pink, white or purple flowers which attract bees and other beneficial insects.

- **HABITAT**
  dry, also suitable for sun

- **SOIL**
  chalk, loam, sand

- **HEIGHT**
  50 cm (20 in)

- **SPREAD**
  50 cm (20 in)

WHERE TO PLANT

The fact that the foliage is scented and the undersides of the leaves are protected with hairs indicate that this is a plant of hot, dry sites and has adapted to a lack of water. Germanders are native Mediterranean plants and suit the same hot, dry conditions as lavender and other shrubby plants. *Teucrium chamaedrys* grows to 50 cm (20 in) tall and so makes a good border edging, especially alongside a path. Try it as ground cover too.

CARING FOR PLANTS

Set out new plants in autumn or spring. Pruning shouldn't be necessary, unless you want to tidy up spent flower heads once they have finished blooming.

THYMUS VULGARIS

Common thyme, garden thyme

ZONE 3-10

Thyme is well known as a culinary herb, but it is also useful in the wider garden. Common thyme is quite variable in size and appearance: it may have pink or white flowers and can be anything between 15 cm (6 in) and 30 cm (12 in) tall. Plenty of other species work well in the garden, too, and nearly all are hardy. The principal requirement for all is full sun and dry, free-draining soil.

- **HABITAT** dry, also suitable for sun
- **SOIL** chalk, loam, sand
- **HEIGHT** 30 cm (12 in)
- **SPREAD** 50 cm (20 in)

WHERE TO PLANT

The tough, wiry stems, tiny leaves and haze of aromatic fragrance mean that thyme can survive hot, dry sites in full sun. It's ideal among paving or stone slabs which reflect the heat back to these sun-loving plants.

CARING FOR PLANTS

Set out new plants in spring or early summer to let them get established before winter. Inevitably, plants get straggly after four or five years, and it is easier to replace rather than try to rejuvenate them, especially as thyme roots easily from cuttings. A light clipping after flowering helps maintain vigor.

ULEX EUROPAEUS
Gorse, furze

See full entry on page 139

VERBASCUM BOMBICYFERUM
Mullein

With its great candelabra of intense yellow flowers, mullein makes a spectacular architectural plant. It grows to about 1.8 m (6 ft) tall, the flower stems rising up from a huge flat rosette of woolly silver leaves. Even the stems and the bracts that encircle the flower buds are smothered in furry, white hairs. The cultivar 'Polarsommer' is similar in size and appearance but has white flowers.

- **HABITAT** dry, also suitable for sun
- **SOIL** chalk, loam
- **HEIGHT** 1.8 m (6 ft)
- **SPREAD** 60 cm (2 ft)

Pictured right

WHERE TO PLANT
With its heavily protected leaves and stems, mullein thrives in hot, dry sites. Its vast basal rosette of leaves offers useful protection from rain and sun on slopes and banks, helping to stabilize the soil surface. It puts up with a wide range of soils.

CARING FOR PLANTS
Mulleins often behave like biennials and die after flowering and are only short-lived as perennials. None of which matters too much as they usually self-seed prolifically. You might need to stake the flower stems in a windy garden.

VIBURNUM TINUS
Laurustinus, viburnum

See full entry on page 83

YUCCA FILAMENTOSA
Adam's needle

See full entry on page 155

DRY

DAMP

ASTILBE CHINENSIS

Astilbe

   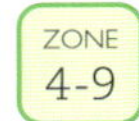

Astilbes have fern-like leaves and plumes of tiny flowers held on erect, wiry stems. Flowers come in shades of pink, red and purple, plus white. Massed together they can create a burst of midsummer color. *Astilbe chinesis* var. *pumila* has flowers in a surprising shade of mauve.

- **HABITAT** damp, also suitable for shade
- **SOIL** loam
- **HEIGHT** 1 m (3 ft)
- **SPREAD** 1 m (3 ft)

WHERE TO PLANT

Astilbes need moisture all year, so be sure of your site before planting. If your damp patch is just a seasonal problem that dries up in summer, so will the astilbe. If the soil is permanently wet, the plants will grow in full sun. Set out young plants in autumn or spring and give them a mulch of well-rotted manure or garden compost to keep the soil extra moist while the plants are getting established.

CARING FOR PLANTS

Occasionally, late frosts can scorch off all the flower buds. If an unseasonable drop in temperature is predicted, protect new buds and leaves with a layer of straw, newspaper or plastic. When a clump develops a bare center, lift and divide it, discarding the central portion.

LEFT Group moisture-loving species such as *Dicentra spectabilis*, Himalayan primrose, ferns and hostas together.

CALTHA PALUSTRIS

Marsh marigold

When in flower, it's easy to see that the marsh marigold is part of the buttercup family. The bright yellow, cup-shaped flowers are produced in spring alongside dark green, glossy leaves. Plants form large clumps after a few years, up to 40 cm (16 in) tall and 45 cm (18 in) across. Some naturally occurring varieties have white flowers, while cultivated forms include 'Flore Pleno', which has double yellow flowers.

- **HABITAT** damp, also suitable for sun
- **SOIL** loam
- **HEIGHT** 40 cm (16 in)
- **SPREAD** 45 cm (18 in)

WHERE TO PLANT

Marsh marigolds are native to boggy areas and are ideal for wet ditches and troughs. They can also be grown in the damp margins of a pond or in water, as long as it is no deeper than about 15 cm (6 in). The soil must be permanently damp if not actually waterlogged.

CARING FOR PLANTS

When clumps outgrow their situation, divide them into smaller portions. Do this either when they have finished flowering or in autumn as growth is slowing down.

CORNUS ALBA

Dogwood

Gardeners with space to play with can plant massed groups of dogwood so that their colorful winter stems really stand out in winter with the low sun behind them. 'Elegantissima' is variegated with deep red stems. 'Sibirica' has the brightest red stems. Look out too for *Cornus sericea* and *C. sanguinea*.

- **HABITAT** damp, also suitable for sun, shade
- **SOIL** clay, loam, sand
- **HEIGHT** 2.5 m (8 ft)
- **SPREAD** 2.5 m (8 ft)

WHERE TO PLANT

These shrubby dogwoods, as opposed to those grown for their flowers and berries, are not at all fussy about soil type and do well in damp spots. You'll get the best colorful winter stems if you plant in full sun, but they will also put on a display in shadier areas.

CARING FOR PLANTS

Let newly planted shrubs settle in for a few years before coppicing – the harsh pruning regime that produces winter color. Cut stems down to the ground in spring for bright stems in winter. To avoid leaving a 'hole' in a small garden, cut down only a third of stems.

DIPSACUS FULLONUM

Teasel

From a flat rosette of bristly leaves, teasels send up great branched flowering stems to 3 m (10 ft) with striking, egg-shaped flowerheads.

Each flower head bristles with spines and is studded with hundreds of tiny flowerlets, usually pale purple. Elegant, spiny bracts (a sort of narrow leaf) curl upward around the whole flower head. Bees find the flowers irresistible, and later on, once seed has set, seed-eating birds flock to the plants.

There is another species, *Dipsacus sativus*, which has larger flower heads with paler flowers and more decorative bracts. It grows to the same height and likes the same growing conditions.

- **HABITAT**
  damp, also suitable for sun

- **SOIL**
  any

- **HEIGHT**
  3 m (10 ft)

- **SPREAD**
  1 m (3 ft)

WHERE TO PLANT

Teasels are native to hedgerows, fields and banks on heavy clay soils. They also grow beside streams and are suitable for damp areas of the garden. They are biennials, so if you are sowing from seed, remember that plants will not mature and flower until the following year. Teasels are ideal for wildlife gardens and for bringing height and structure to any plot. It's best not to site them too close to a path or a children's play area because of their prickles.

CARING FOR PLANTS

Biennials usually die after flowering, but teasels are prolific self-seeders, so you shouldn't run short of plants. Teasel seed heads make popular dried flowers but be careful of the spines when cutting them.

ENKIANTHUS CAMPANULATUS
Enkianthus

Enkianthus campanulatus is the hardiest species in this genus of shrubs from the Far East. It flowers in spring with pretty, pendulous, bell-shaped blooms, which may be cream, red or pink, often with contrasting veining on the petals. In autumn the dull green leaves change color to a breathtaking display of reds, yellows and oranges.

- **HABITAT** damp, also suitable for sun, shade
- **SOIL** clay, loam, sand. Acidic
- **HEIGHT** 3 m (10 ft)
- **SPREAD** 3 m (10 ft)

WHERE TO PLANT

Enkianthus does best in moist acidic soil, although it will tolerate neutral soil, especially if lots of leaf mold is dug in before planting. It can put up with quite a bit of shade or will grow in full sun. What it does not like is strong winds. The shrubs are slow growing, reaching a mere 60–90 cm (2–3 ft) after around five years, although eventually they can get to 3 m (10 ft), less tall in colder areas.

Mix with other plants that thrive in acidic soil, such as rhododendrons, pieris and azaleas, for an interesting contrast of foliage.

CARING FOR PLANTS

Mulch with leaf mold in spring. As it blooms on the previous year's wood, prune only to remove dead or damaged branches. If necessary, prune immediately after flowering.

FILIPENDULA ULMARIA

Meadowsweet

Meadowsweet, a familiar wild flower of damp meadows and ditches with its froth of creamy, sweet-smelling plumes in summer, has also been bred for the garden. *Filipendula ulmaria* 'Rosea' has pale pink flowers, and 'Variegata' has yellow-variegated leaves. The plants have blood-red stalks that contrast with the ribbed, almost pleated green leaves.

Two other vigorous species enjoy similar conditions. *F. purpurea* has plumes of purplish flowers. *F. rubra* 'Venusta' has deep pink flowers and grows to about 2 m (6½ ft) tall.

- **HABITAT**
  damp, also suitable for shade

- **SOIL**
  clay, loam

- **HEIGHT**
  2 m (6 ½ft)

- **SPREAD**
  1 m (3 ft)

WHERE TO PLANT

Match meadowsweet's natural habitat and use it to plant up permanently wet patches in the garden, especially areas in semi-shade. It suits damp ditches and bog gardens and will grow in the wet margins of ponds.

CARING FOR PLANTS

Set plants out in late winter or spring. Overgrown clumps can be divided at the same time of year. If the foliage gets unsightly towards the end of the year, cut back to ground level.

Some gardeners grow meadowsweet for its foliage alone and will snip off flowers to avoid detracting from the leaves – especially with 'Variegata'.

KALMIA LATIFOLIA

Calico bush, mountain laurel

The flower buds of the calico bush look like perfect pink sugar cake decorations, and when they open they retain a curious crimped appearance. The bush flowers in late spring and early summer, and for the rest of the year it is a handsome evergreen, densely packed with glossy, oval leaves.

- **HABITAT** damp, also suitable for sun, shade
- **SOIL** clay, loam, sand. Acidic
- **HEIGHT** 3 m (10 ft)
- **SPREAD** 3 m (10 ft)

WHERE TO PLANT

The calico bush makes an interesting companion to rhododendrons in acidic soil. Plant it in a woodland setting under taller trees – it grows to a maximum 3 m (10 ft) – where it will do well in the shade, but will flower more freely the sunnier the position.

CARING FOR PLANTS

Deadhead regularly to keep the bush looking good while it is in flower. Pruning shouldn't be necessary, but if the shrub gets out of shape or there are one or two untidy branches cut these out in spring. The calico bush flowers on the previous year's wood, so any pruning must be done straight after flowering. A small new bush may take a few years to flower.

KIRENGESHOMA PALMATA

Kirengeshoma

An unusual and elegant plant from the Far East with lobed leaves rather like an oriental maple. In late summer it sends up spires of creamy-white flowers, each with a neatly contrasting dark brown calyx capping the petals. Plants can vary in height, ranging from 60 cm to 1.2 m (2 to 4 ft).

- **HABITAT**
  damp, also suitable for shade

- **SOIL**
  clay, loam, sand. Acidic

- **HEIGHT**
  1.2 m (4 ft)

- **SPREAD**
  1 m (3 ft)

WHERE TO PLANT

In its native habitat in Japan, *Kirengeshoma palmata* is a tough perennial woodland plant and can be used to underplant shrubs and trees, preferably where the soil is moist.
It prefers acidic soil but may cope in neutral conditions if other factors are favorable and plenty of leaf mold is worked into the soil before planting.

CARING FOR PLANTS

If plants show signs of flagging in dry spells, water them thoroughly. A mulch of garden compost or leaf mold will help to conserve moisture.

LIGULARIA MACROPHYLLA

Ligularia

The yellow, daisy-like flowers of *Ligularia macrophylla* are massed together on tall spires. The flowering stems can reach 1.5 m (5 ft) and make a bold planting in midsummer. Equally prominent are its leaves: gray-green and roughly oval, they can be up to 60 cm (2 ft) long.

- **HABITAT**
  damp, also suitable for shade

- **SOIL**
  clay, loam

- **HEIGHT**
  1.5 m (5 ft)

- **SPREAD**
  1 m (3 ft)

WHERE TO PLANT

Ligularia is a waterside plant, growing best when its roots are in soil that is permanently wet. Plant in the boggy margins of a pond, where it can tolerate a little sun. Alternatively, chose a site that is moist rather than wet, but make sure the spot is permanently shady. Give it plenty of room to show off its huge, handsome leaves.

CARING FOR PLANTS

In an exposed garden the flower stems may need staking. Strong winds can also cause the leaves to wilt. Give plants a thick mulch every spring with well-rotted garden compost.

MOLINIA CAERULEA *SUBSP.* CAERULEA

Purple moor grass

     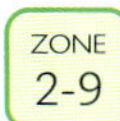

Ornamental moor grasses contribute delicate flowers and valuable autumn color to the garden. *Molinia caerulea* subsp. *caerulea*, one of the lower-growing species, forms clumps that are a manageable 1 m (3 ft) high. It has typical, grass-like, purple flowers in late summer, and the leaves turn butter yellow in the autumn. There are many cultivars, including 'Moorhexe', which is slightly smaller, and 'Moorflamme', which is the deepest purple of all.

- **HABITAT** damp, also suitable for exposed
- **SOIL** acidic
- **HEIGHT** 1 m (3 ft)
- **SPREAD** 1 m (3 ft)

WHERE TO PLANT

Purple moor grass is native to acidic peat bogs, so plant it in any acidic, permanently boggy ground. It will also do well in merely damp soil, as long as it doesn't dry out. As it is a plant of open, exposed sites, it can cope with full sun.

CARING FOR PLANTS

Molinia is one of the few ornamental grasses that can be slow to establish. One way around this is to start off with the largest plants you can afford. It's also unusual in that its leaves become detached and blow away in late autumn – most deciduous grasses need a sharp tug to detach dead leaves.

PERSICARIA BISTORTA

Bistort, snakeweed

ZONE 4-9

The densely packed spikes of small pink flowers of bistort look best massed together in a great drift. It has strong 'knotted' stems and elongated, grass-like leaves. The name snakeweed refers to its twisted roots.

The genus is a versatile one. *Persicaria amphibia* will grow in more than 30 cm (12 in) of water. *P. affinis* makes low-growing ground cover in sun or light shade. *P. campanulata* has fragrant looser flower heads arranged in clusters and does well in clay.

- **HABITAT** damp, also suitable for sun, shade

- **SOIL** any

- **HEIGHT** 60 cm (2 ft)

- **SPREAD** 1 m (3 ft)

WHERE TO PLANT

P. bistorta is a plant of damp meadows and water margins. Use its upright flower stems, which can be to 60 cm (2 ft) or so, to add strong vertical lines to plantings beside ponds, in a mixed border with damp soil, or to brighten up the shade in an edge-of-woodland setting.

CARING FOR PLANTS

If bistort finds conditions too much to its liking, it can become invasive. When it threatens to overwhelm neighboring plants, divide and shift clumps in autumn or spring. You may prefer to remove spent flower heads before they turn an unsightly brown.

PRIMULA FLORINDAE
Giant cowslip

      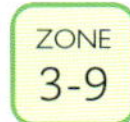

Giant cowslip is native to the Himalayas and is suitably large. Its typical primrose-type leaves are extra-long, forming a rosette to 1 m (3 ft) across, from which it sends up strong stems topped with clusters of yellow flowers like the diminutive wild primula. The flowers are sweetly scented and the stems may reach a height of about 1.2 m (4 ft).

- **HABITAT**
  damp, also suitable for sun, shade

- **SOIL**
  loam

- **HEIGHT**
  1.2 m (4 ft)

- **SPREAD**
  1 m (3 ft)

WHERE TO PLANT

Mimic giant cowslip's natural cool, streamside habitat by planting it in ground that never dries out – in a bog garden, for example, or in the margins of a pond. Provided the soil stays moist, the cowslip can put up with full sun, although shade is more to its liking. It also tolerates acidic soil (although it does not actually need it) and grows well in heavy clay.

CARING FOR PLANTS

Leave flowers to set seed and you may be rewarded with self-sown seedlings if conditions are right. Do not let newly established plants dry out.

RHEUM PALMATUM

Chinese rhubarb

ZONE 6-10

The height of the flowering stems – 2.5 m (8 ft) – means that Chinese rhubarb makes a big impact in a garden. Its flowers are fine and feathery and reddish-pink, but it is the leaves that are the true stars. They are huge, deeply lobed and hairy; those of the cultivar 'Atrosanguineum' are purplish-red when young.

- **HABITAT** damp, also suitable for sun, shade
- **SOIL** clay, loam
- **HEIGHT** 2.5 m (8 ft)
- **SPREAD** 2.5 m (8 ft)

WHERE TO PLANT

Turn a bog garden into a feature with Chinese rhubarb You could plant one at the edge of a large pond – it would dwarf a small one – and it doesn't actually have to grow in water as long as the soil does not dry out. Mulching will help.

CARING FOR PLANTS

Removing flower stems before they get a chance to bloom results in the plant putting all its energies into producing even bigger leaves. Plants appreciate a mulch of well-rotted manure or home-made garden compost in spring.

RODGERSIA PINNATA

Rodgersia

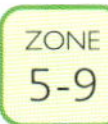 ZONE 5-9

Rodgersias are grown for their beautiful architectural foliage. The leaves, to 1 m (3 ft) long, are tinged with bronze when they first appear, fading to dark green. Their bold shapes, coupled with their height, stand out against less well-defined foliage and give a border an almost tropical look. Plants do not always flower reliably, but when they do, they send up plumes of tiny, starry flowers in shades of pink, cream or white.

- **HABITAT**
  damp, also suitable for sun, shade

- **SOIL**
  clay, loam. Acidic

- **HEIGHT**
  1.5 m (5 ft)

- **SPREAD**
  1.5 m (5 ft)

WHERE TO PLANT

Water is the most important element. A bog garden is ideal, but rodgersias will grow in any naturally moist, water-retentive soil in sun or shade. Choose a sheltered spot protected from the coldest winds.

CARING FOR PLANTS

Do not let plants dry out. Plant in spring or autumn. Dig in plenty of well-rotted manure or garden compost beforehand and mulch regularly throughout the year to help keep the soil moist.

SALIX

Willow

Willows come in all shapes and sizes from shrubs to the huge weeping species. Shrub-sized species are best for average gardens, though you can use stooling as a technique to keep larger ones within bounds. Grow them for their elegant foliage or colorful stems. *Salix elaeagnus* has pretty silver leaves; *S. purpurea* 'Nancy Saunders' has dark purple stems.

- **HABITAT**
  damp, also suitable for sun, coastal

- **SOIL**
  clay, loam, sand

- **HEIGHT**
  5 m (16 ft)

- **SPREAD**
  4 m (13 ft)

WHERE TO PLANT

Anywhere where boggy soil is a problem. They are easy to establish from whips (small rooted plants) or cuttings pushed straight into the ground. Use them to make windbreaks and hedges.

CARING FOR PLANTS

To keep a large willow in check, cut it right down to the ground alternate years in late winter (stooling).

EXPOSED

BERBERIS THUNBERGII

Barberry

   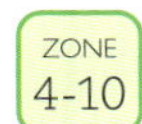

ZONE
4-10

There are both evergreen and deciduous species in this genus and the deciduous ones, such as *Berberis thunbergii*, have intensely colored foliage in autumn. The leaves of *B. atropurpurea*, for example, are tinged with purple, then turn a vivid red before they drop. At the same time you should get a colorful crop of small, shiny, red berries.

- **HABITAT**
  exposed, also suitable for sun

- **SOIL**
  any

- **HEIGHT**
  1.5 m (5 ft)

- **SPREAD**
  1.5 m (5 ft)

WHERE TO PLANT

Barberry is a dense, thorny shrub that provides shelter for more delicate plants. It also makes a tough property hedge, especially as it eventually grows to 1.5 m (5 ft) or more. When planting as a hedge or windbreak, set out young plants about 45 cm (18 in) apart and put up a temporary screen, such as some netting or trellis, to protect them until the roots get well established. Cut off between a third and a quarter of the top growth to encourage them to bush out.

CARING FOR PLANTS

Barberries can be grown as an informal or a formal hedge, left to spread out or clipped into a neat shape. To keep an informal hedge full of healthy, strong shoots, take out any very old or straggly stems at the end of winter before the shrubs come into flower or later on, in summer, after flowering.

LEFT Low-growing shrubby plants such as heathers and smaller shrubs like some species of hebes that form low, domed shapes can cope with exposed sites

EXPOSED

CRATAEGUS MONOGYNA

Common hawthorn, quickthorn

Hawthorn is most often grown as a hedge, although left unpruned it will develop into a small tree. It has small, lobed leaves and sprays of powerfully scented white flowers in spring. These are followed by red berries (haws) in autumn. For a red-flowered version, try *Crataegus laevigata* 'Paul's Scarlet'; 'Rosea Flore Pleno' has double pink flowers. Both are just as tough as the single-flowered form.

WHERE TO PLANT

Hawthorn is densely branched and spiny, making it ideal for a boundary hedge on an exposed site. Or grow it as a specimen tree in an open, sunny site, where it may top 10 m (30 ft) after twenty years or so.

In a wildlife garden, hawthorns will provide safe nesting sites for birds in spring and a good food supply of berries later in the year.

CARING FOR PLANTS

To make an impenetrable hedge, plant bare-root saplings in winter, setting them 30 cm (1 ft) apart and place a second, staggered row in front. To keep a hedge in check, clip it after flowering if possible, although winter trimming will do.

Hawthorn is prone to fireblight. Prune out and burn all dead, or damaged, branches at least 25–30 cm (10–12 in) below the damaged area every year. Avoid excessive nitrogen fertilizer.

- **HABITAT** exposed, also suitable for sun, coastal
- **SOIL** any
- **HEIGHT** 10 m (33 ft)
- **SPREAD** 10 m (33 ft)

Pictured right

EXPOSED

ELAEAGNUS PUNGENS

Elaeagnus

This large, evergreen shrub is grown for its foliage. The dark green leaves have a curious metallic sheen on the upper surface and dull silver tinge beneath. There are several cultivars. 'Goldrim' has leaves with pronounced gold edges; the popular 'Maculata' has bold gold splashes on the leaves.

Although it is grown as a foliage plant, it has tiny fragrant autumn flowers hidden deep between the leaves. If you're not in the know, it's a puzzle to work out where the perfume is coming from. Very occasionally, flowers go on to set fruit. The berries produced can be red, brown or even silver.

WHERE TO PLANT

The shrub is not fussy about soil and is not damaged by strong winds, even salt-laden coastal gales. Use it as hedging – on its own or in a mixed hedge – to create a windbreak. After some years it can get to 4 m (13 ft) high. Elaeagnus also tolerates partial shade: the variegated cultivars are useful for cheering up a dull corner.

CARING FOR PLANTS

Set out new plants in spring or autumn. Pruning shouldn't be necessary, except to keep shrubs in shape. If you plant variegated cultivars, cut out any shoots that revert to plain green.

- **HABITAT**
  exposed, also suitable for sun, shade, dry, coastal
- **SOIL**
  any
- **HEIGHT**
  4 m (13 ft)
- **SPREAD**
  4 m (13 ft)

ERICA CINEREA

Heath, bell heather

Heathers make undemanding plants once established and *Erica cinerea* is just one of hundreds. The species has spires of white, intense pink or purple flowers in summer and autumn, and its dark green leaves are evergreen. Its spreading habit makes it a useful ground-cover plant. You may need to use a mulch, such as bark chippings, to keep weeds down until the plants spread.

WHERE TO PLANT

Plant heathers out in the open, away from deciduous trees: dead leaves snagged in among the heather is not a good look, especially as *E. cinerea* is autumn flowering. An exposed windy site is fine – just imagine the windswept moors where they grow naturally – but heathers must have acidic soil.

CARING FOR PLANTS

Set out new plants in late autumn, if possible when the soil is still warm. Water plants in hot spells for the first year or two, and they will develop far-reaching, healthy root systems that spare you the trouble of further nurturing in years to come. Clip lightly after flowering to maintain plants in good, compact shape.

- **HABITAT**
  exposed, also suitable for sun, dry

- **SOIL**
  clay, loam, sand. Acidic

- **HEIGHT**
  45 cm (18 in)

- **SPREAD**
  1 m (3 ft)

ESCALLONIA

Escallonia

Escallonias are generally evergreen, with glossy leaves and small but pretty clusters of pink or red flowers throughout summer and on into autumn. 'Donard Seedling', one of the taller cultivars, grows eventually to about 3 m (10 ft). There are a number of cultivars, but some are slightly tender, so choose one of the more vigorous ones for an exposed garden.

WHERE TO PLANT

Escallonias not only survive strong winds unshaken but can also tolerate salt-laden gales along coasts. For this reason they make excellent hedging for exposed gardens and provide valuable shelter for less tolerant plants.

CARING FOR PLANTS

Severe weather in spring can be a setback when the plants are putting on new growth, but the hardier escallonias are usually tough enough to fight back. Any dieback simply acts as a sort of natural pruning and the plants swiftly recover. Pruning is necessary only to remove unsightly scorching or to reshape an unruly shrub.

- **HABITAT**
  exposed, also suitable for sun, coastal

- **SOIL**
  any

- **HEIGHT**
  3 m (10 ft)

- **SPREAD**
  3 m (10 ft)

EUONYMUS FORTUNEI

Euonymus

No one grows euonymus for the flowers; the foliage is everything. 'Emerald 'n' Gold' has butter-yellow leaves with a central splash of strong green; in winter the leaves develop a pink edging. 'Emerald Gaiety' has white-edged leaves that turn bronze in the autumn.

- **HABITAT**
  exposed, also suitable for shade, dry, coastal

- **SOIL**
  any

- **HEIGHT**
  1 m (3 ft)

- **SPREAD**
  1.5 m (5 ft)

WHERE TO PLANT

Use 'Emerald 'n' Gold' to cheer up barren, shady borders along the side of a house or garage. Although it will climb, you could train it up (or down) to cover a steep, shady slope. It grows well in chalk soils, and in heavy clay to a slightly lesser extent.

CARING FOR PLANTS

Pruning is not strictly necessary, although older, untidy plants do respond well to being cut back hard at any time of the year except spring.

If any stems of variegated cultivars revert to green, cut them out immediately.

FUCHSIA MAGELLANICA

Magellan fuchsia

ZONE
6-9

There are hundreds of cultivars of fuchsia with elaborate frilled flowers in extraordinary color combinations, but for tough hedging and windbreaks only the hardy species, *Fuchsia magellanica*, will do. It has classic, pendant flowers in a rather ecclesiastical combination of purple and red, followed by small, purple, sausage-shaped fruits.

- **HABITAT**
  exposed, also suitable for shade, coastal

- **SOIL**
  any

- **HEIGHT**
  2.5 m (8 ft)

- **SPREAD**
  1.5 m (5 ft)

WHERE TO PLANT

Hardy fuchsias grow wild on mild British coastlines, rearing up into great shaggy hedges and dwarfing low cottages tucked into the landscape. In the garden they make effective windbreaks and attractive informal hedges. In mild areas this species may remain green all winter long, but even when it loses its leaves the twiggy framework still filters strong winds.

CARING FOR PLANTS

Cut back new plants to about 30 cm (12 in) before their first winter. This will stop the wind from rocking the plants and unsettling new root systems. If you have planted fuchsias late in the summer, give them a little frost protection – bark chips or newspaper tucked around the stems. Even though all growth above ground will be killed by a severe winter, this species will grow back quickly from the base.

GAILLARDIA

Blanketflower

Gaillardia, also known as blanketflower, is a native of the prairies of North America. It has big bold flowers rather like sunflowers, with a central boss of stamens surrounded by two-tone petals in bright yellows and reds. Its common name is said to come from Native American blankets whose colors the flowers echo. Flowers are produced all summer long and well into autumn.

- **HABITAT** exposed, also suitable for dry
- **SOIL** any well drained
- **HEIGHT** 1 m (3 ft)
- **SPREAD** 50 cm (20 in)

WHERE TO PLANT

Choose a sunny border and keep in mind that some varieties grow up to 1 m (3 ft) tall. Once established they form strong sturdy plants that shouldn't need propping up. Perennial varieties tend to be short-lived, especially if soil gets waterlogged, so if you garden on heavy soil and crave gaillardia, try growing them in big pots or create a raised border.

CARING FOR PLANTS

Gaillardia can be annuals or perennials. Grow annuals from seed, starting them off indoors and then hardening off before setting them in their final positions in the border. Like many summer flowers, rich soil promotes leaf growth rather than blooms, so don't be tempted to give them a dose of fertilizer. Deadheading will keep the flowers coming but remember to allow some to set seed if you want to increase stocks next year.

HEBE ALBICANS
Hebe

ZONE 8-10

Hebe albicans is a pretty, fairly low-growing shrub that seldom exceeds 1 m (3 ft). It makes a satisfying dome shape as it grows. The small, gray-green, oval leaves are thick and fleshy and offset small spires of white flowers in late spring and summer.

- **HABITAT** exposed, also suitable for sun, coastal

- **SOIL** any

- **HEIGHT** 1 m (3 ft)

- **SPREAD** 1 m (3 ft)

WHERE TO PLANT

This hebe does well in most soils and is particularly useful in sandy or chalky soil. The grayish foliage helps it stand up to full sun, and its compact dome shape survives in coastal sites, where it is exposed to salt-laden winds. Most hebes do well in similar situations, and the genus is large so you can choose from taller species, which are ideal for hedging, to smaller plants, which will thrive in containers. One thing to watch out for is hardiness: *H. albicans* is hardy to -10°C (14°F) for a short period, but some species are relatively tender.

CARING FOR PLANTS

Set out new plants in autumn or spring. Old bushes that have become unsightly and straggly can be cut back hard in spring, but not too hard. Leave some older branches to help start the shrub back into life. These can be cut out the following spring.

HIPPOPHAE RHAMNOIDES

Sea buckthorn

See full entry on page 148

KNIPHOFIA CAULESCENS

Red-hot poker, torch lily, poker plant

ZONE 6-9

Great, knotted, knobby stems of red-hot pokers lie along the ground and send up exotic rosettes of gray-green leaves. They flower late in summer and on into autumn. The flower stems gave rise to their common name: the newest florets at the top of spikes are an intense red as they open, fading to yellow further down as they age.

WHERE TO PLANT

Choose a spot in full sunshine in fairly poor soil. Set them at the back of a border, not only because of their height, which can top 1.2 m (4 ft) or more, but so you can plant in front of them to disguise their unprepossessing appearance early in the season.

There are about seventy species of red-hot poker, but *Kniphofia caulescens* is the only truly hardy one. If your garden is cold, it's probably a waste of time experimenting with others, apart from perhaps *K. uvaria*, which is relatively hardy and has bright red flowers. Adding a protective winter mulch when the foliage has died back will help.

CARING FOR PLANTS

If an exceptionally cold spell is forecast, it may be wise to add a layer of branches, wood chips or newspaper to protect overwintering plants. For best results, get new plants in the garden in early autumn or wait until spring.

- **HABITAT** exposed, also suitable for sun, dry, coastal
- **SOIL** loam, sand
- **HEIGHT** 1.2 m (4 ft)
- **SPREAD** 1 m (3 ft)

Pictured left

LIMONIUM PLATYPHYLLUM

Sea lavender, statice

See full entry on page 150

MOLINIA CAERULEA SUBSP. CAERULEA

Purple moor grass

See full entry on page 118

NEPETA X FAASSENII

Catmint

See full entry on page 98

ONOPORDUM ACANTHIUM

Scotch thistle, cotton thistle

See full entry on page 99

EXPOSED

ROSA RUGOSA

Hedgehog rose, Ramanas rose

Rosa rugosa has simple, open, flat flowers of deepest pink with a central mass of yellow stamens. It is a very tough, vigorous, hardy rose, which stays in bloom from summer to autumn, after which it forms big, bold, orangey-red hips. Even the leaves contribute to the color scheme, turning golden-yellow before they drop.

- **HABITAT**
  exposed, also suitable for sun, coastal

- **SOIL**
  any

- **HEIGHT**
  2.5 m (8 ft)

- **SPREAD**
  2.5 m (8 ft)

WHERE TO PLANT

Its dense, thorny habit makes it an ideal boundary hedge, shrugging off cold winds and heavy or salt-laden gales. Depending on conditions, it will grow to 1.2–2.5 m (4–8 ft) tall. Set plants out 45 cm (18 in) apart for an effective hedge or windbreak.

CARING FOR PLANTS

When growing R. *rugosa* as a hedge, prune plants hard in their first year. Clip and tidy up as necessary. Dig in plenty of well-rotted manure before planting and feed plants regularly.

STIPA GIGANTEA

Giant feather grass, golden oats

Giant feather grass forms a dense clump of dark leaves about 75 cm (30 in) tall. In summer, spectacular flowering stems shoot up to 2.5 m (8 ft), bearing spikes of typical grassy flowers. These change color as they age, from golden to straw yellow to reddish-brown. The plant makes an impact all year, and only a heavy snowfall will finally see off old flower stems.

For a scaled-down option in a smaller garden, plant *Stipa tenuissima* (Mexican feather grass). It makes eye-catching ground cover and is extraordinarily mobile, moving gracefully in the slightest breeze. It will also self-seed.

- **HABITAT**
  exposed, also suitable for sun, dry

- **SOIL**
  chalk, loam, sand

- **HEIGHT**
  2.5 m (8 ft)

- **SPREAD**
  1 m (3 ft)

WHERE TO PLANT

Grasses from the genus *Stipa* are native to wide open, dry grasslands throughout the world. Set them in full sun in the driest part of the garden. They do best in poor soil: rich soil tends to encourage lax, floppy growth.

CARING FOR PLANTS

Cut out dead flower stems of S. *gigantea* in winter or shear right across the clump, but don't cut it right back to the ground in winter. Wait until mid-spring before carrying out drastic treatment. Pull out spent stems of S. *tenuissima* in late winter

ULEX EUROPAEUS

Gorse, furze

This shrub, which is native to western and central Europe, has dark green stems, few leaves and vicious spines. It flowers first in spring and intermittently throughout the summer with typical pea-like, yellow flowers that give off a delicious scent of honey on hot days. A mature shrub can reach 2.5 m (8 ft) tall.

- **HABITAT** exposed, also suitable for sun, dry, coastal
- **SOIL** sand
- **HEIGHT** 2.5 m (8 ft)
- **SPREAD** 2.5 m (8 ft)

Pictured left

WHERE TO PLANT

Its almost universal tolerance of difficult conditions makes it surprising that gorse isn't more widely planted, until you consider its extremely spiny nature. This means that gorse cannot really be grown as anything other than a windbreak or an effective, impenetrable property hedge. Plant it on awkward, dry, sunny banks or use it to shelter more delicate species in an exposed garden. Do not plant it close to paths, for obvious reasons.

CARING FOR PLANTS

Tidy up gorse bushes after the first flush of flowers in spring. Occasionally, late frosts will scorch new shoots, but these can be easily trimmed back.

VERBENA BONARIENSIS

Argentinian verbena

See full entry on page 59

LEFT Take your cue from nature: gorse flourishes on exposed seaward sites. It flowers nearly all year round in mild areas.

"

COASTAL

BRACHYGLOTTIS MONROI

Monro's ragwort

ZONE 7-10

A small, rounded shrub that puts up with fierce sun and regular wind. It reaches about 1 m (3 ft) tall and has thickly felted gray-green leaves that are white underneath.

WHERE TO PLANT
Plant a row of brachyglottis in spring to provide shelter for more vulnerable plants. Set them out about 45 cm (18 in) apart and protect with a temporary windbreak in the first year or two until the roots take a firm hold. Thereafter they will do the job for you. If you want to use brachyglottis as a privacy hedge, plant Dunedin Group 'Sunshine', which grows to 1.5 m (5 ft) tall.

CARING FOR PLANTS
Trim back bushes lightly after flowering to tidy them up or prune them more severely for a formal appearance. For an extra flush of new growth, cut back plants hard in spring.

- **HABITAT** coastal, also suitable for dry, exposed

- **SOIL** chalk, loam, sand

- **HEIGHT** 1 m (3 ft)

- **SPREAD** 1.5 m (5 ft)

LEFT Driftwood fencing filters the wind and a gravel mulch retains vital moisture in a coastal garden.

BUPLEURUM FRUTICOSUM

Shrubby hare's ear

An often overlooked but useful shrub with glossy, blue-green, evergreen leaves and studded with masses of tiny, star-shaped, yellow-green flowers all through the summer and into autumn.

- **HABITAT**
  coastal, also suitable for dry, exposed

- **SOIL**
  chalk, loam, sand

- **HEIGHT**
  1.5 m (5 ft)

- **SPREAD**
  1.5 m (5 ft)

WHERE TO PLANT

The tough, leathery leaves of bupleurum resist the drying effects of wind and sun and salt-laden winds. (These same properties also help it resist polluted city atmospheres, where grit-laden winds can damage less well-armoured leaves.) Because it is initially low-growing and spreads quickly, bupleurum is useful ground cover, effectively choking weeds. It will grow to 1.5 m (5 ft) if left unpruned.

CARING FOR PLANTS

Cut away any untidy shoots in mid- to late spring and any stems that have suffered scorching or the effects of a severe winter. To get a flush of completely fresh new growth and to keep plants small, prune hard at this time of year.

CENTRANTHUS RUBER

Red valerian

The leaves of red valerian have a grayish tinge that hints at their tolerance of hot, sunny places, as does their thickened, rather fleshy appearance. In late spring and summer and on into autumn plants bear clusters of tiny, red, nectar-rich flowers, which attract butterflies and moths. There is also a white-flowered cultivar, 'Albus', and others in various shades of red-pink.

- **HABITAT** coastal, also suitable for sun
- **SOIL** chalk, loam, sand
- **HEIGHT** 1 m (3 ft)
- **SPREAD** 50 cm (20 in)

WHERE TO PLANT

Red valerian is a common sight on limestone cliffs by the sea and translates easily into similar garden situations. It puts up with chalky and sandy soils and can find a foothold in crevices in old walls or paving. If you offer it a better position in a border, it generally does less well, producing floppy growth and fewer flowers.

CARING FOR PLANTS

Cutting the flowers – either after they have finished or for arranging in water – encourages more flowers to form. At the end of the season, leave some to set seed for a stock of future plants.

CHOISYA TERNATA
Mexican orange blossom

See full entry on page 90

CISTUS x CYPRIUS

Rock rose, sun rose

This hybrid rock rose has large, papery, white flowers in summer. Each petal is splashed with a blotch of deep purplish-red at the base, which highlights a boss of yellow stamens. The leaves have a strong balsamic fragrance, which is released on hot days or if you rub a leaf between your fingers. It spreads rapidly to form a nicely rounded shrub, which can ultimately be about 1.5 m (5 ft) tall and across. Rock roses are not reliably hardy, and they are often fairly short-lived.

- **HABITAT**
  coastal, also suitable for sun, dry

- **SOIL**
  any

- **HEIGHT**
  1.5 m (5 ft)

- **SPREAD**
  1.5 m (5 ft)

WHERE TO PLANT

Rock roses grow wild all over the Mediterranean and are adapted to harsh, sunny climates. Plant *Cistus* x *cyprius* on open, sunny ground, in poor or sandy soil, and it will flourish. Its height when fully grown helps to protect smaller plants by acting as a windbreak in exposed gardens. Get new plants into the ground in spring so they are well established by winter.

CARING FOR PLANTS

Deadheading will keep a rock rose flowering for longer as well as improve its appearance. It shouldn't need pruning, but any dead or spindly branches can be cut out in spring. Take cuttings in summer as a safeguard against plant loss in severe winters.

CRAMBE MARITIMA

Sea kale

    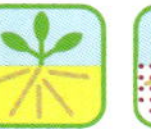   

Sea kale has large, blue-green leaves, crinkled and twisted just like vegetable kale. In early summer plants send up flower stems around 1 m (3 ft) tall, topped with a mass of tiny white flowers, sweetly honey-scented. *Crambe cordifolia* is a cultivated garden species. It is much taller – to 2.5 m (8 ft) – but still suitable for sunny coastal gardens. In bloom, its huge, airy flowerheads make a haze of petals and fine, thin stems.

- **HABITAT** coastal, also suitable for sun, dry, exposed
- **SOIL** sand, poor, stony
- **HEIGHT** 2.5 m (8 ft)
- **SPREAD** 1 m (3 ft)

WHERE TO PLANT

In the wild, sea kale sends up shoots on beaches above the tide line. In the garden put it in a gravel bed, on poor, stony soil, or in free-draining, sandy soil. The leaves die back after flowering quite early on in the summer, so make sure there are some neighboring plants that can spread out and cover up gaps.

CARING FOR PLANTS

Give sea kale plants plenty of space, setting them at least 1 m (3 ft) apart in spring. Keep an eye out for slugs when seedlings are still tender and vulnerable. Older plants can also be divided at this time of year. Sea kale flowers early in the summer. Sow seed under glass in spring, then plant out the seedlings when they are strong enough to survive outside.

CRATAEGUS MONOGYNA

Common hawthorn, quickthorn

See full entry on page 126

CYTISUS SCOPARIUS

Common broom

A common shrub in its native Western Europe. It has tough, wiry, green stems, and the leaves are reduced to tiny leaflets perfectly adapted to conserve water. By late spring it is smothered in pretty yellow flowers, similar in shape (but not size) to sweet peas. These are followed by typical pea pod seed cases, and on a hot day you can actually hear them popping as they burst open and scatter seeds.

- **HABITAT**
  coastal, also suitable for sun, dry

- **SOIL**
  loam, sand

- **HEIGHT**
  1.5 m (5 ft)

- **SPREAD**
  1.5 m (5 ft)

WHERE TO PLANT

Common broom grows to about 1.5 m (5 ft) high and makes a good shelter or specimen bush in difficult conditions. It grows naturally in hot, open sites where the ground is poor and baked dry in the summer, and it prefers free-draining, sandy soil that does not become waterlogged. The white-flowered Portuguese broom (*Cytisus multiflorus*) tolerates similar conditions.

CARING FOR PLANTS

Brooms tend to be short-lived. When they start to look straggly and leggy, be prepared and have some replacements waiting in the wings. The best time to take cuttings is late summer. Take them with a heel – a strip of the main stem to which they were attached – for best results.

Brooms do not really need pruning, but if you need to keep a bush in check, trim it back after flowering. Never cut into old wood because it will not regenerate.

ELAEAGNUS PUNGENS
Elaeagnus

See full entry on page 128

ERYNGIUM BOURGATII

Sea holly

ZONE 5-9

The thistle-like flower heads of sea holly are produced in summer, studded with tiny blue flowers that match the plant's unusual blue stems. Shields of spiny bracts radiate from the flower heads. The flowers are especially attractive to bees, and as they age their color changes in intensity from steely blue to a paler lavender.

- **HABITAT** coastal, also suitable for sun, dry

- **SOIL** poor

- **HEIGHT** 45 cm (18 in)

- **SPREAD** 45 cm (18 in)

WHERE TO PLANT

Eryngium bourgatii is well adapted to thrive in hot, dry gardens. It doesn't mind sandy soils, and its tough, silver-green leaves resist the drying action of winds. Sea holly plants should reach at least 45 cm (18 in) tall when in flower and spread to make a good-sized clump – another wind- and drought-resistant feature of plants that have adapted to tough conditions.

CARING FOR PLANTS

Eryngiums must have well-drained soil or they can be prone to rot in wet winters. In a prolonged cold, wet spell, clear away any dead leaves from the base of the plant to prevent moisture lingering. They do best in poor soil: rich soil promotes soft, floppy growth.

HIPPOPHAE RHAMNOIDES

Sea buckthorn

Native to large areas of Europe and temperate Asia, this shrub has made its way into the gardener's repertoire because of its resilience to harsh conditions. The linear leaves are covered in silvery scales for protection against sun scorch and water loss. Female and male flowers are borne on different plants, and to be sure of a crop of orange autumn berries, you need to plant one male bush with several females.

WHERE TO PLANT

In the wild, sea buckthorn grows on hot, dry, free-draining sand dunes. In the garden give it the hottest, driest site you can. Its tough nature makes it perfect for coastal gardens, and its thorniness makes it a useful hedge. It grows to around 3 m (10 ft) tall, producing a tangled mass of branches that just keep on spreading. Although deciduous, it is so densely branched that it makes an excellent windbreak even in winter.

CARING FOR PLANTS

If you need to reshape a bush or are using sea buckthorn in a more formal hedge, do this in summer, but otherwise pruning shouldn't be necessary.

- **HABITAT**
  coastal, also suitable for sun, exposed

- **SOIL**
  loam, sand

- **HEIGHT**
  3 m (10 ft)

- **SPREAD**
  3 m (10 ft)

Pictured right

COASTAL

LIMONIUM PLATYPHYLLUM

Sea lavender, statice

        ZONE 3-9

The tough, wiry stems and leathery leaves of sea lavender are typical of a plant evolved to withstand difficult coastal conditions. The leaves are arranged in a rosette, which helps to conserve moisture and stabilize the soil. In summer and autumn it's a haze of tiny, pale lavender flowers, which are a rich source of nectar for bees and butterflies.

- **HABITAT** coastal, also suitable for sun, dry, exposed
- **SOIL** chalk, loam, sand
- **HEIGHT** 1 m (3 ft)
- **SPREAD** 50 cm (20 in)

WHERE TO PLANT

In a coastal garden, plant sea lavender in borders or alongside paths, wherever you like. Inland, plant it in gravel beds or in hot, open spaces baked by the sun. Take advantage of its airy flower heads to create a pale purple cloud that will be a foil for bolder blooms.

CARING FOR PLANTS

Set out new plants in spring. Deadhead flowers to keep plants blooming or simply cut stems for the house. Cut out all flowering stems once the plant has finished flowering.

NEPETA X FAASSENII
Catmint

See full entry on page 98

PEROVSKIA ATRIPLICIFOLIA
Russian sage

See full entry on page 53

PHLOMIS FRUTICOSA

Jerusalem sage

In summer the subtle silver-gray foliage of Jerusalem sage is offset by whorls of two-lipped flowers in a refined shade of mustard yellow. The flowers are long-lasting, and after the petals fall, the seed heads are worth retaining for their interesting shape.

- **HABITAT**
  coastal, also suitable for sun, dry

- **SOIL**
  any

- **HEIGHT**
  1 m (3 ft)

- **SPREAD**
  1.5 m (5 ft)

WHERE TO PLANT

Plant in a hot, dry spot in full sun: it will thrive in drought conditions. Although Jerusalem sage probably does best in free-draining chalk or sandy soils, it will still perform in clay as long as it doesn't get waterlogged.

CARING FOR PLANTS

Cutting the whole bush back hard in spring rejuvenates the foliage, which can get a bit dusty-looking after a long hot summer, but this will delay flowering by a few weeks.

ROSA RUGOSA
Hedgehog rose, Ramanas rose

See full entry on page 136

SKIMMIA JAPONICA

Japanese skimmia

Sprays of pinkish-red flower buds make an attractive display all through autumn and winter before finally opening in spring. The flowers within are white and sweetly scented. Male and female flowers are carried on separate plants, and both sexes are needed for the female plants to bear bright red summer berries.

- **HABITAT** coastal, also suitable for shade, dry

- **SOIL** chalk, loam, sand

- **HEIGHT** 1.5 m (5 ft)

- **SPREAD** 1.5 m (5 ft)

WHERE TO PLANT

Skimmia japonica grows in soil that ranges from acidic to neutral. Truly versatile, it does equally well in dry or damp soils, in sun or shade. Skimmias will grow in coastal and city gardens because of the marginally milder micro-climates. Plant among taller trees and shrubs for added protection against frost.

CARING FOR PLANTS

Severe frost can damage new growth, which will turn white. Cut out any affected stems in late spring. Shrubs naturally form a rounded dome, ultimately around 1.5 m (5 ft) tall. The only pruning necessary is to take out any shoots that spoil the shape.

TAMARIX RAMOSISSIMA

Tamarisk

A familiar sight in many of Europe's coastal towns, in public spaces, in gardens, and right on the sea walls where they are regularly drenched with spray at high tide. The overall impression they make is feathery, from the plumy sprays of pink flowers to the finely divided foliage. *Tamarix ramosissima* has pink flowers in late summer and autumn. It reaches about 6 m (20 ft) high. The other commonly grown species, *T. tetrandra*, is smaller – 4 m (13 ft) – and flowers earlier.

- **HABITAT**
  coastal, also suitable for sun, dry
- **SOIL**
  loam, sand
- **HEIGHT**
  6 m (20 ft)
- **SPREAD**
  6 m (20 ft)

WHERE TO PLANT

In exposed coastal gardens tamarisk makes a useful shelter and a tough boundary hedge that is completely resilient to salt-laden gales. Sandy soils present no problem, and the bushes can withstand prolonged dry spells. They will not thrive in heavy clay soils that become waterlogged in winter.

CARING FOR PLANTS

To thicken up bushes of *T. ramosissima* for a windbreak or hedge, prune them hard once they start growing in spring. Pruning also prevents plants from becoming top-heavy, making them unstable and susceptible to wind rock.

ULEX EUROPAEUS
Gorse, furze

See full entry on page 139

VIBURNUM TINUS
Laurustinus, viburnum

See full entry on page 83

YUCCA FILAMENTOSA

Adam's needle

ZONE 4-10

The yucca plant's long, sword-like leaves can be up to 75 cm (30 in) long and sprout from a central rosette. In late summer these plants, which are native to the southeastern states of the USA, send up tall flowering spikes, to 2 m (6½ ft) tall, packed with nodding flowers, like pendent inverted tulips, usually white with a hint of cream or green. The flowers last several weeks before fading. Look out for the cultivars 'Bright Edge', which has broad yellow margins to the leaves, and 'Variegata', with white edges.

WHERE TO PLANT

Make a feature of such a stunning specimen plant in a hot, dry spot or in a gravelly coastal garden. A yucca can look larger than life and dwarf any surrounding plants, so is probably best placed where it can be admired without distraction or where you need a vertical accent.

CARING FOR PLANTS

Cut down the flower stems when they have finished blooming. Plants in warmer areas should be fine over winter, but in frostier areas a winter mulch is wise, though this species is the hardiest of the genus and does quite well. Do not be fooled by its warm origins – desert nights can be very cold.

- **HABITAT**
  coastal, also suitable for sun, dry

- **SOIL**
  chalk, loam, sand

- **HEIGHT**
  2 m (6½ ft)

- **SPREAD**
  1.5 m (5 ft)

Pictured left

PICTURE CREDITS

All images from Shutterstock

Cover © 2017 Wiert nieuman; p2–3 © 2017 Alexey Lobanov; p4–5 © 2013 jlynx; p6 © 2020 Bildagentur Zoonar GmbH; p8 © 2020 SergaFomin; p10 © 2018 guentermanaus; p11 © Dzm1try; p13 top © 2017 Anna Gratys; p13 bottom © 2014 Elena Elisseeva; p14 © 2018 ata_moro; p15 © 2020 Tanja Esser; p16 © 2018 Maria Evseyeva; p17 © 2016 Christopher Lyzcen; p18 © 2017 Wojmac; p19 top © 2018 Peter Turner Photography; p19 bottom © 2018 Ollga P; p20 © 2016 asadykov; p21 © 2015 1000 Words; p22 © 2018 Martchan; p23 © 2018 bluejava1; p24 © 2017 Kathryn Roach; p25 © 2018 Svitlyk; p26 left © 2020 Alexandra Kovaleva; p26 right © 2018 Wiert nieuman; p27 © 2015 bymandesigns; p28 © 2011 DUSAN ZIDAR; p29 © 2018 Mintr; p30 © 2020 Veronika Fialova; p31 © 2020 AVN Photo Lab; p32–33 © 2017 Katarzyna Mazurowska; p34–35 © 2017 Wiert nieuman; p36 © 2017 BENCHA STEWART; p38 © 2015 Linda Hughes Photography; p39 © 2017 nnattalli; p40 © 2017 speakingtomato; p41 © 2018 Alan Kean; p42 © 2018 Halit Omer; p43 © 2017 Dirk M. de Boer; p44 © 2013 Mariusz S. Jurgielewicz; p45 © 2012 Lijuan Guo; p46 © 2018 Audrey Wilson1; p48 © 2022 Nahhana; p49 © 2017 Gardens by Design; p51 © 2016 dadalia; p52 © 2019 Ken Schulze; p53 © 2013 loflo69; p54 © 2018 Alphabetman; p56 © 2018 Federico Magonio; p57 © 2019 galsand; p58 © 2014 dailin; p60 © 2015 Hannamariah; p61 © 2010 wiwsphotos; p62 © 2019 Nancy J. Ondra; p63 © 2016 simona pavan; p64 © 2017 Peter Turner Photography; p65 © 2010 godrick; p66 © 2018 nnattalli; p67 © 2019 Furiarossa; p68 © 2018 Moiseenko Maksim; p69 © 2017 IanRedding; p70 © 2016 fotomarekka; p71 © 2016 WindOfHope; p72 © 2019 Noel V. Baebler; p73 © 2014 Flegere; p74 © 2018 Natalia van D; p76 © 2019 Gerry Bishop; p78 © 2019 Victoria Tucholka; p79 © 2019 Svitlyk; p80 © 2021 Alex Manders; p81 © 2020 Peter Turner Photography; p82 © 2020 Orest lyzhechka; p83 © 2019 Roel Meijer; p84 © 2017 Thoreau; p85 © 2019 zzz555zzz; p86 © 2018 Jaiz Anuar; p87 © 2018 Peter Turner Photography; p88 © 2018 simona pavan; p89 © 2017 Iva Vagnerova; p90 © 2018 nnattalli; p91 © 2013 Yulia Grigoryeva; p93 © 2018 Klaus Wagenhaeuser; p94 © 2017 photonkova; p96 © 2014 flaviano fabrizi, p97 © 2019 Gabriela Beres; p98 © 2017 Anna Gratys; p99 © 2019 COULANGES; p100 © 2017 Roman Khomlyak; p101 © 2018 KanphotoSS; p102 © 2016 Deniza 40x; p103 © 2020 Alex Manders; p104 © 2020 Orest lyzhechka; p105 © 2018 Diyana Dimitrova; p107 © 2020 encierro; p108 © 2017 Gardens by Design; p109 © 2016 guentermanaus; p110 © 2016 Jakub Koziol; p111 © 2016 Peter Turner Photography; p112 © 2018 Veroja; p113 © 2019 JIANG TIANMU; p114 © 2017 Starover Sibiriak; p115 © 2017 Rita Robinson; p116 © 2018 V.Borisov; p117 © 2017 Nadzeya Pakhomava; p118 © 2016 guentermanaus; p119 © 2019 Gabriela Beres; p120 © 2021 Olga Glagazina; p121 © 2018 c_WaldWiese; p122 © 2017 Fotych; p123 © 2020 Ihor Hvozdetskyi; p124 © 2013 steve estvanik; p125 © 2018 Grisha Bruev; p127 © 2013 AlessandroZocc; p128 © 2018 simona pavan; p129 top © 2018 arousa; p129 bottom © 2018 Nahhana; p130 © 2018 Hecos; p131 © 2018 chaistock; p132 © 2020 NataFrank; p133 © 2018 c_WaldWiese; p134 © 2017 BabichAndrew; p136 © 2019 Przemyslaw Muszynski; p137 © 2018 Peter Turner Photography; p138 © 2020 Gianluca Piccin; p139 © 2012 Graeme Dawes; p140 © 2016 Gardens by Design, p141 © 2018 Max_555; p142 © 2020 Bildagentur Zoonar GmbH; p143 © 2015 Philip Bird LRPS CPAGB; p144 © 2018 Steve Photography; p145 © 2018 Yvonne60; p146 © 2019 seaonweb; p147 © 2019 JohnatAPW; p149 © 2012 loskutnikov; p150 © 2016 IanRedding; p151 © 2019 Matauw; p152 © 2018 Anita van den Broek; p153 © 2021 Flower_Garden; p154 © 2019 aniana Pg155 © 2022 Maliflower73